Advances in Information Security

Volume 76

Series editor
Sushil Jajodia, George Mason University, Fairfax, VA, USA

More information about this series at http://www.springer.com/series/5576

Suryadipta Majumdar • Taous Madi • Yushun Wang
Azadeh Tabiban • Momen Oqaily
Amir Alimohammadifar • Yosr Jarraya
Makan Pourzandi • Lingyu Wang
Mourad Debbabi

Cloud Security Auditing

 Springer

Suryadipta Majumdar
Information Security and Digital Forensi
University at Albany - SUNY
Albany, NY, USA

Taous Madi
CIISE
Concordia University
Montréal, QC, Canada

Yushun Wang
CIISE
Concordia University
Montreal, QC, Canada

Azadeh Tabiban
CIISE
Concordia University
Montreal, QC, Canada

Momen Oqaily
CIISE
Concordia University
Montreal, QC, Canada

Amir Alimohammadifar
Concordia University
Montreal, QC, Canada

Yosr Jarraya
Ericsson Security Research
Saint-Laurent, QC, Canada

Makan Pourzandi
Ericsson Security Research
Montreal, QC, Canada

Lingyu Wang ⓘD
CIISE
Concordia University
Montreal, QC, Canada

Mourad Debbabi
Concordia University
Montreal, QC, Canada

ISSN 1568-2633
Advances in Information Security
ISBN 978-3-030-23130-9   ISBN 978-3-030-23128-6   (eBook)
https://doi.org/10.1007/978-3-030-23128-6

This Springer imprint is published by the registered company Springer Nature Switzerland AG.
The registered company address is: Gewerbestrasse 11, 6330 Cham, Switzerland

Preface

Auditing and security assurance verification is, and has traditionally been, essential for many industries due to regulatory requirements. Examples of this are ISO 27001 for information security management, the Health Insurance Portability and Accountability Act (HIPAA), and the Payment Card Industry Data Security Standard (PCI DSS). With the current deployment of cloud computing technologies in IT solutions and mobile communication networks comes the need to extend auditing and security assurance verification to the cloud.

The cloud, due to its elastic, on-demand, and self-service nature, is attracting many new applications from different fields, making it part of the cyber critical infrastructure and part of everyday life. Cloud computing technologies such as virtualization allow physical resources to be dynamically shared through virtual instances, and in the latest generation of mobile communication networks the network function virtualization approach instantiates virtual functions to deliver network services for different verticals.

However, while the unique properties of the cloud (elastic, self-service, multi-tenancy) generate opportunities for high efficiency and customization, they also bring novel security challenges. Perhaps most importantly, multi-tenancy allows different customers to share physical resources, opening the door to possible attacks through virtual layers. Also, the increased complexity due to multiple abstraction layers in the cloud (such as virtual and physical layers covering different resources for networking, compute, and storage) may cause inconsistencies which need auditing mechanisms spanning through all different layers. While in classical auditing we had only two actors (service provider and user), the cloud setup introduces a new actor: the cloud service provider. The auditing therefore needs to account for the addition of new actors. Lastly, the dynamic nature of the cloud means changing the entire virtual resource configuration at runtime, to scale in and out in response to service load. This in turn may invalidate earlier auditing results and calls for a continuous and adaptive auditing approach.

In short, the complex design and implementation of a cloud infrastructure may cause vulnerabilities and misconfigurations, which adversaries could exploit to conduct serious damage. The security concerns and challenges therefore generate a demand for transparency and accountability, and the auditing becomes a main element in building the confidence of cloud tenants and users in their cloud providers. Because auditing of services in the cloud ecosystem differs in many ways from existing auditing approaches, new methods are required to span larger and deeper at different virtual and physical layers, to account for new actors, and to provide continuous and adaptive security auditing.

This book is the result of a collaboration between academic and industrial researchers to address real-world issues in the cloud environment. It addresses the major challenges and presents new auditing methods for the cloud. Pointing to the problem encountered when auditing in the cloud, the authors cover security threats from all cloud abstraction levels and propose auditing mechanisms for a wide range of security and functional requirements from the cloud tenants and users to the relevant security standards.

Through their journey, the authors first tackle the complexity of the cloud due to its virtual and physical layers, concentrating on the most challenging aspects of virtual L2–L3 networks. Essential issues for auditing in the cloud are considered: from user-level auditing verification to virtual network isolation verification. For user-level auditing, a method is presented verifying an extensive list of security properties related to authentication and authorization mechanisms in the cloud. The security property verification is then extended to the virtual network layer. The virtual network layer 2 auditing addresses various security concerns generated from potential inconsistencies between cloud layers and tenant isolation breaches while the virtual layer 3 auditing tackles the threats from multi-tenancy.

Next, the authors consider the need for an efficient continuous auditing approach to address the dynamic nature of the cloud. They propose three different auditing methods: retroactive, intercept-and-check, and proactive. Retroactive auditing subscribes to the traditional way of auditing by identifying security breaches after the fact. The intercept-and-check auditing enables continuous auditing, specifically elaborating on enabling runtime security policy enforcement. Finally, the proactive auditing addresses the need to efficiently audit the cloud at runtime. At the end, the authors propose a runtime security policy enforcement in a widely used open-source cloud platform as an approach to continuous security auditing in the cloud.

This book provides a comprehensive state-of-the-art knowledge on cloud security auditing, covering a wide range of security auditing issues and proposing new solutions to the challenges. It is beneficial to researchers and engineers involved in security for cloud computing.

Stockholm, Sweden   Eva Fogelström
March 2019

Acknowledgments

We would like to express our deepest gratitude to all people who contributed to the realization of this work. This book is a result of research work supported by the Natural Sciences and Engineering Research Council of Canada and Ericsson Canada under CRD Grants (N01566 and N01823) and by PROMPT Quebec.

Contents

Chapter 1
Introduction

Cloud computing has been gaining momentum as a promising IT solution specially for enabling ubiquitous, convenient, and on-demand accesses to a shared pool of configurable computing resources. Businesses of all sizes nowadays leverage cloud services for conducting their major operations (e.g., web service, inventory management, customer service, etc.). Based on the way services are provided, cloud computing has been divided into different categories such as infrastructure as a service (IaaS), platform as a service (PaaS), and software as a service (SaaS). In most of these categories, there exist at least three main stakeholders: cloud service providers, tenants and their users. A cloud service provider owns a significant amount of computational resources, e.g., servers, storage, and networking, and offers different paid services (e.g., IaaS, PaaS, etc.) to its customers by utilizing this pool of resources. Usually, cloud tenants are different companies or departments within a company. A tenant, the direct customer of cloud providers, enjoys the ad hoc and elastic (i.e., allocating/deprovisioning based on demands) nature of cloud in utilizing the shared pool of resources for conducting its necessary operations. As a member of a cloud tenant, a user mainly avails different services offered by a tenant. Thus, by providing a dynamic (i.e., ever changing) and a measured service (i.e., "pay as you go") to its users and tenants, cloud computing has become a popular choice for diverse business models in recent years.

While cloud computing has seen such increasing interests and adoption, the fear of losing control and governance among its tenants still persists due to the lack of transparency and trust. In order to guarantee that their desired security policies are respected in the cloud, there is a need for auditing and verifying the security of cloud against such policies. In this book, we discuss how state-of-the-art security auditing solutions may help to increase cloud tenants' trust in the service providers by providing assurance on the compliance with the applicable laws, regulations, policies, and standards. This book covers basic to advanced (e.g., retroactive, runtime, and proactive) auditing techniques to serve different stakeholders in the cloud and to build the trust and transparency between cloud tenants and service

© Springer Nature Switzerland AG 2019
S. Majumdar et al., *Cloud Security Auditing*, Advances in Information Security 76,
https://doi.org/10.1007/978-3-030-23128-6_1

providers. This book also discusses a wide range of security properties related to cloud-specific standards (e.g., Cloud Control Matrix (CCM) and ISO 27017). Finally, this book elaborates on the integration of security auditing solutions into modern cloud management platforms (e.g., OpenStack, Amazon AWS, and Google GCP).

1.1 Motivations

Cloud service providers typically adopt the multi-tenancy model to optimize resources usage and achieve the promised cost-effectiveness. Multi-tenancy in the cloud is a double-edged sword. While it enables cost-effective resource sharing, it increases security risks for the hosted applications. Indeed, multiplexing virtual resources belonging to different tenants on the same physical substrate may lead to critical security concerns such as cross-tenants data leakage and denial of service.

To overcome these concerns, security auditing and compliance validation might be a promising solution. However, there are currently many challenges in the area of cloud auditing and compliance validation. For instance, there exists a significant gap between the high-level recommendations provided in most cloud-specific standards (e.g., Cloud Control Matrix (CCM) [17] and ISO 27017 [52]) and the low-level logging information currently available in existing cloud infrastructures (e.g., OpenStack [90]). In practice, only limited forms of auditing may be performed by cloud subscriber administrators [81], and there exist a few automated compliance tools (e.g., [28, 112]) with several major limitations, which are discussed later in this section.

Furthermore, the unique characteristics of cloud computing may introduce additional complexity to the auditing task. For instance, the use of heterogeneous solutions for deploying cloud systems may complicate the data collection and processing step in auditing. Moreover, the sheer scale of a cloud, together with its self-provisioning, elastic, and dynamic nature, may render the overhead of many verification techniques prohibitive. In particular, the multi-tenant and self-service nature of clouds usually imply significant operational complexity, which may prepare the floor for misconfigurations and vulnerabilities leading to violations of security compliance. Therefore, the security compliance verification with respect to security standards, policies, and properties is desirable to both cloud providers, and its tenants and users. Evidently, the Cloud Security Alliance (CSA) has recently introduced the Security, Trust and Assurance Registry (STAR) for security assurance in clouds, which defines three levels of certifications (self-auditing, third-party auditing, and continuous, near real-time verification of security compliance) [19]. However, the above-mentioned complexities coupled with the sheer size of clouds (e.g., a decent-sized cloud is said to have around 1000 tenants and 100,000 users [92]) imply one of the main challenges in cloud security auditing. In summary, the major challenges are to handle the unique nature of clouds and to deal with their sheer size in providing a scalable and efficient security auditing solution for clouds.

To this end, existing approaches can be roughly divided into three categories (a more detailed review of related works will be given in Chap. 2). First, the retroactive approaches (e.g., [28, 112]) catch compliance violations after the fact by verifying different configurations and logs of the cloud. However, they cannot prevent security breaches from propagating or causing potentially irreversible damages (e.g., leaks of confidential information or denial of service). Second, the intercept-and-check approaches (e.g., [13, 89]) verify the compliance of each user request before either granting or denying it, which may lead to a substantial delay in responding each user request. Third, the proactive approaches in [13, 89] verify future change plan to identify any potential breach from the proposed plan. However, due to the dynamic and ad hoc nature of clouds, providing future change plan in advance is not always feasible, and hence, this approach is not practical for clouds. In summary, existing works suffer from at least one of the following limitations: (1) supporting a very limited set of security properties, (2) responding with a significant delay, and (3) involving unrealistic amounts of manual efforts. This book introduces solutions to address these above-mentioned limitations.

1.2   Cloud Security Auditing

Though security auditing is not a new process, automation of this process and complexity of targeted infrastructures introduce non-trivial challenges. Based on the proposed solutions and best practices, we identify different phases (as in Fig. 1.1) of an automated security auditing process.

Defining the Scope and Threat Model As a very first step, an organization should define the scope of its auditing. Part of it is to identify the critical and sensitive assets, operations, and the modules in the system that deal with those assets and operations. The following step is to identify threats or nature of threats to be considered for the auditing process. Most of the time, threat model depends on the nature of the business and demand of customers. Part of this step is to

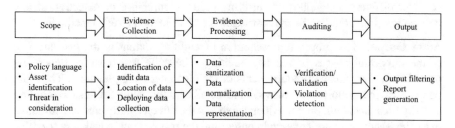

Fig. 1.1 Different steps of the cloud security auditing process

describe security assumptions while considering each threat. To this end, in last few years, different studies have been conducted to identify risks and threats in the cloud computing ecosystem. Based on those threats, several security properties are proposed by CSA [15], ENISA [31], ISO [52], NIST [77], CUMULUS [21], etc.

Data/Evidence Collection The next phase is to gather evidences/data to conduct the audit process. Based on the target system and threat model, audit data is listed. In some cases (e.g., cloud and distributed systems), locating those audit data is non-trivial. Also, the data collection phase has become more dynamic with the virtualization and multi-tenancy, which results in an increase in the amount of data to be collected. In this book, we also consider security aspects of data collection in addition to the different runtime and continuous data collection techniques of different data types. The trust model ensures that the audit data provided by a tenant is real and fresh. At the same time, there might exist privacy concerns in a central auditing system, e.g., tenant must not leak any sensitive information to the auditor, and no information should be leaked across different tenants.

In the cloud, most of the audit data are events, logs, and system configurations. Data collection techniques vary in terms of targeted environments and data, e.g., what data to collect based on the scope, threat model, and objectives, and how to collect data (more challenging in a cloud-based system).

Data/Evidence Processing The collected raw data from the previous step requires further processing to be able to conduct auditing. In case of verifying compliance with a policy language, it depends on the language. As data is collected from different sources, the collected data needs to be sanitized. For better understanding and interpretation, different correlation methods are applied on sanitized data to categorize them. There are several techniques (e.g., call graph [67], information flow graph [11], and reachability graph [117]) to represent the audit data. Heterogeneous data is normalized by different methods, e.g., [26]. Storing this processed audit data is also an important phase especially when dynamic cloud auditing generates an enormous amount of data over time.

Auditing In the auditing phase, processed data is verified against the policies for any violation. The process either validates the system or detects an anomaly if it exists. There are many auditing techniques proposed over time, though comparatively fewer automated techniques exist for the cloud. In this book, we consider different techniques of verifying policy compliance or detection of any policy violation including formal verification and validation (V&V) methods.

Audit Output The proper representation of auditing output is the last and one of the important phases of security auditing. The audit report varies depending on different demands and requirements of the customers (e.g., tenants). The main purpose of this report is to provide the details of the audit findings (including security breaches) with proper evidences so that the affected party can take initiative to fix the issues. A major concern in outputting the result is not to leak any sensitive and unnecessary information to any tenant [117]. Proper information isolation must be ensured.

1.3   Challenges in Cloud Security Auditing

Even though auditing has been in practice for years, the unique nature, such as dynamic, elastic, self-service, of clouds brings new challenges in devising security auditing solutions in the cloud environment. In the following, we discuss the key challenges in cloud security auditing.

High-Level Security Properties There is a significant gap between the high-level standards (defining security requirements) and the low-level cloud configurations (providing the audit data). Even though several works (e.g., [63–65]) highlight this challenge and partially address the concern, the issue still persists in interpreting security guidelines and defining security properties ready to be used in auditing solutions. Current solutions rely on manual identification of security properties, which is infeasible and error-prone, especially when we consider multiple layers of the cloud and intend to provide a unified security solution.

Non-Trivial Log Processing One mandatory and non-trivial step of cloud security auditing is log processing. This step involves several challenging tasks. First, identifying the heterogeneous sources of audit data requires a good understanding of the deployed cloud system, which usually consists of several complex components, e.g., management platform and layer-2 plugins. Second, due to the different ways for storing (e.g., database and text files) the configurations and logs, the collection of audit data has to be performed with different methods. Finally, the diverse format of the logs requires extensive processing efforts to unify and integrate different formats before using the results in auditing.

Reducing Manual Involvement Automating the auditing process is a must in a dynamic environment like cloud to ensure the accuracy and efficiency. However, the current solutions still rely on manual efforts in several critical steps. Fully eliminating or at least reducing manual effort is not trivial mainly for the following two reasons. First, defining the security properties is a mandatory step for any auditing process and existing works mostly rely on human inputs for this step. Existing rule mining techniques in access control might be useful in automating this step. Second, most auditing approaches (as reported in Chap. 2) rely on manual identification of security properties and critical operations (which potentially can violate a property). Applying machine learning or more specifically interactive machine learning techniques may reduce the manual efforts involved in this step.

Runtime Policy Enforcement The traditional auditing mechanism, which is conducted retroactively, is not always useful for a dynamic environment like cloud. In other words, the state of the cloud usually changes frequently, and thus, a retroactive auditing approach cannot stop any irreversible damages to the cloud. Furthermore, the self-allocation and deprovisioning of resources invalidate the auditing results very often. Therefore, in today's clouds, we need runtime and continuous auditing solutions. However, the sheer size of cloud may create a significant obstacle in providing such a runtime auditing mechanism.

Practical Response Time Most of the existing runtime auditing solutions (further elaborated in Chap. 2) suffer from a significant delay, and thus, these solutions become infeasible for real-world deployment. In this book, we carefully investigate this shortcoming, and identify one of the main reasons is that all the auditing steps in these solutions are performed at a single point (a.k.a. critical event[1]) and not taking any advantage of the dependency relationship (more discussed in Chap. 6). Alternatively, leveraging dependencies (if any) among cloud events potentially may allow starting the auditing process in advance and conducting auditing incrementally.

1.4 Outline

The remainder of the book is organized as follows:

* Chapter 2 describes the existing works on cloud security auditing. To this end, we categorize existing solutions into three classes. First, the retroactive auditing approach verifies security policies after the fact. Second, the intercept-and-check auditing solutions intercept each management operation in the cloud and verify the impact of any change before applying it on the cloud. Third, the proactive auditing approach verifies security policies in advance and enforces the policy at runtime.
* Chapter 3 discusses an automated framework that allows auditing the cloud infrastructure from the structural point of view while focusing on virtualization-related security properties and consistency between multiple control layers. Furthermore, to show the feasibility of our approach, we integrate our auditing system into OpenStack, one of the most used cloud infrastructure management systems. To show the scalability and validity of our framework, we present our experimental results on assessing several properties related to auditing inter-layer consistency, virtual machines co-residence, and virtual resources isolation.
* Chapter 4 describes an off-line automated framework for auditing consistent isolation between virtual networks in OpenStack-managed cloud spanning over overlay and layer 2 by considering both cloud layers' views. To capture the semantics of the audited data and its relation to consistent isolation requirement, we devise a multi-layered model for data related to each cloud-stack layer's view. Furthermore, we integrate our auditing system into OpenStack, and present our experimental results on assessing several properties related to virtual network isolation and consistency. Our results show that our approach can be successfully used to detect virtual network isolation breaches for large OpenStack-based data centers in a reasonable time.

---

[1]The event type which potentially can violate a security property.

- Chapter 5 presents a runtime security auditing framework for the cloud with special focus on the user level including common access control and authentication mechanisms e.g., RBAC, ABAC, and SSO. We implement and evaluate the framework based on OpenStack, a widely deployed cloud management system. The main idea towards reducing the response time to a practical level is to perform the costly operations for only once, which is followed by significantly more efficient incremental runtime verification.
- Chapter 6 describes a learning-based proactive security auditing system. To this end, we design a stand-alone log processor for clouds, which may potentially be used for various log analyses. Consequently, we leverage the log processor outputs to extract probabilistic dependencies from runtime events for the dependency models. Finally, through these dependency models, we proactively prepare for security critical events and prevent security violations resulting from those critical events.
- Chapter 7 elaborates the design and implementation of a middleware as a pluggable interface to OpenStack for intercepting and verifying the legitimacy of user requests at runtime while leveraging our previous work on proactive security verification to improve the efficiency. We describe in detail the middleware implementation and demonstrate its usefulness through a use case.

Chapter 2
Literature Review

This chapter first categorizes the existing cloud security auditing, then elaborates each category mainly based on its coverage and adopted verification techniques, and finally presents a taxonomy based on these works. There exist mainly three categories of cloud security auditing approaches. In the following, we discuss each of these approaches with corresponding example works.

2.1 Retroactive Auditing

Retroactive auditing approach (e.g., [9, 27, 28, 56, 63, 65, 112, 116, 118]) in the cloud is a traditional way to verify the compliance of different components of a cloud. Works under this approach in the cloud target a wide range of security properties that cover various cloud layers, such as data, user, and virtual infrastructure.

There are several works that target auditing data location and storage in the cloud (e.g., [50, 56, 116, 118]). Wang et al. [116] propose a cloud storage system which enables privacy-friendly public auditing to ensure data security in the proposed system. The work leverages public key based homomorphic linear authenticator (HLA) to significantly reduce the communication and computation overhead at the auditor side. Kai et al. [56] can handle multiple auditing requests to verify the data integrity in the multi-cloud environment. In addition, similar to the former this work preserves the privacy of the audit data. On the other hand, Ismail et al. [50] propose a game theory based auditing approach to verify the compliance of data backup requirements of users. Unlike previous ones, Wang et al. [118] offer auditing of data origin and consistency in addition to data integrity.

There exist other works, which target virtual infrastructure change auditing (e.g., [27, 28, 63–65, 112]). These works cover different layers (e.g., user, virtual network, etc.) in the virtual infrastructure. Particularly, Ullah et al. [112] propose an architecture to build an automated security compliance tool for cloud computing

© Springer Nature Switzerland AG 2019
S. Majumdar et al., *Cloud Security Auditing*, Advances in Information Security 76,
https://doi.org/10.1007/978-3-030-23128-6_2

platforms focusing on auditing clock synchronization and remote administrative and diagnostic port protection. Doelitzscher [27] proposes on-demand audit architecture for IaaS clouds and an implementation based on software agents to enable anomaly detection systems to identify anomalies in IaaS clouds for the purpose of auditing. The works in [27, 112] have the same general objective, which is cloud auditing, but they use empirical techniques to perform auditing, whereas we use formal techniques to model and solve the auditing problem. Madi et al. [63, 64] verify a list of security properties to audit the cross-layer consistencies in the cloud.

In addition, several industrial efforts include solutions to support cloud auditing in specific cloud environments. For instance, Microsoft proposes SecGuru [9] to audit Azure datacenter network policy using the SMT solver Z3. IBM also provides a set of monitoring tool integrated with QRadar [48], which is a security information and event management system, to collect and analyze events in the cloud. Amazon is offering web API logs and metric data to their AWS clients by AWS CloudWatch and CloudTrail [4] that could be used for the auditing purpose. Although those efforts may potentially assist auditing tasks, none of them directly supports auditing a wide range of security properties covering authentication, authorization, and virtual infrastructure on cloud standards.

Furthermore, there are several auditing solutions (e.g., [38, 41, 42, 65, 110]) targeting the user level (e.g., authentication and authorization) of the cloud. Majumdar et al. [65] verify the role-based access control implementation in OpenStack, a popular cloud platform. This work also verifies a list of security properties to ensure proper implementation of authentication steps in the cloud. To accommodate the need of secure collaborative environments such as cloud computing, there have been some efforts towards proposing multi-domain/multi-tenant access control models (e.g., [38, 41, 110]). Gouglidis and Mavridis [41] leverage graph theory algorithms to verify a subset of the access control security properties. Gouglidis et al. [42] utilize model checking to verify custom extensions of RBAC with multi-domains [41] against security properties. Lu et al. [61] use set theory to formalize policy conflicts in the context of inter-operation in the multi-domain environment.

2.2   Intercept-and-Check Auditing

Existing intercept-and-check approaches (e.g., [13, 46, 62, 69, 89, 105, 107]) perform major verification tasks while holding the event instances blocked. Works under this category cover the virtual network, user level, and software-defined network (SDN) layers of a cloud environment as discussed in the following.

The works (e.g., [13, 89]) at the virtual network level are mainly verifying the security properties to safeguard multiple layers in a virtual network through an intercept-and-check approach. These works focus on operational network properties (e.g., black holes and forwarding loops) in virtual networks, whereas our effort is oriented towards preserving compliance with structural security properties that impact isolation in cloud virtualized infrastructures. Designing cloud monitoring services based on security service-level agreements have been discussed in [96].

The user-level runtime auditing is proposed in Patron [62] and Majumdar et al. [69]. More specifically, Patron [62] audits the access control rules defined by the cloud tenants. In addition, Patron enforces these rules on the cloud by leveraging the middleware supported in OpenStack, one of the major cloud platforms. Majumdar et al. [69] utilize similar interception approach in OpenStack and audit the proper deployment of various authentication and authorization plugins, such as single sign-on (SSO), role-based access control (RBAC), and attribute-based access control (ABAC) in the cloud.

There are also few works (e.g., TopoGuard [46] and TopoGuard+ [105]) which adopt the intercept-and-check approach in the software-defined network (SDN) environment. TopoGuard [46] and TopoGuard+ [105] perform the interception and enforcement to prevent topology tempering attacks in SDN. Those works in SDN can be complements to the above-mentioned solutions for other layers in the cloud.

2.3 Proactive Auditing

The concept of proactive security auditing for clouds is different than the traditional security auditing concept. The first proactive auditing approach for clouds is proposed in [13]. Additionally, the Cloud Security Alliance (CSA) recommends continuous auditing as the highest level of auditing [19], from which latter works (e.g., [66, 67]) are inspired. The current proactive and runtime auditing mechanisms are more of a combination of traditional auditing and incident management. For example, LeaPS [67] learns from incidents and intercepted events to process or detect in a similar manner as a traditional incident management system. At the same time, LeaPS verifies and enforces compliance against different security properties, which are mostly taken from different security standards, and provide detailed evidence for any violation through LeaPS dashboard. Therefore, the concept of proactive security auditing is a combination of incident management and security auditing.

Proactive security analysis has also been explored for software security enforcement through monitoring programs' behaviors and taking specific actions (e.g., warning) in case security policies are violated. Many state-based formal models are proposed for those program monitors over the last two decades. First, Schneider [103] modeled program monitors using an infinite-state-automata model to enforce safety properties. Those automata recognize invalid behaviors and halt the target application before the violation occurs. Ligatti [59] builds on Schneider's model and defines a more general program monitors model based on the so-called edit/security automata. Rather than just recognizing executions, edit automata-based monitors are able to suppress bad and/or insert new actions, hence transforming invalid executions into valid ones. Mandatory Result Automata (MRA) is another model proposed by Ligatti et al. [29, 60] that can transform both actions and results to valid ones. Narain [75] proactively generates correct network configurations using the model finder Alloy, which leverages a state-of-the-art SAT solver. To this end, they specify a set of end-to-end requirements in first order logic and determine the set of existing network components. Alloy uses a state-of-the-art SAT solver to provide the configurations that satisfy the input requirements for each network component. Considering the huge size of cloud environments and the tremendous space of possible events, adapting those solutions in the cloud is possibly very challenging.

Weatherman [13] is aiming at mitigating misconfigurations and enforcing security policies in a virtualized infrastructure. Weatherman has both online and offline approaches. Their online approach intercepts management operations for analysis, and relays them to the management hosts only if Weatherman confirms no security violation caused by those operations. Otherwise, they are rejected with an error signal to the requester. The work defines a realization model that captures the virtualized infrastructure configuration and topology in a graph-based model. The latter is synchronized with the actual infrastructure using the approach in [12]. Two major limitations of this proposition are: (1) the model capturing the whole infrastructure causes a scalability issue for the solution and (2) the time consuming operation-checking that should be performed on the emergence of each event makes security enforcement not feasible for large size data centers. Congress [89] is an OpenStack project offering both online and offline policy enforcement approaches. The offline approach requires submitting a future change plan to Congress, so that the changes can be simulated and the impacts of those changes can be verified against specific properties. In the online approach, Congress first applies the operation to the cloud, then checks its impacts. In case of a violation, the operation is reverted. However, the time elapsed before reverting the operation can be critical to perform some illicit actions, for instance, transferring sensitive files before losing the assigned role. Foley et al. [34] provide an algebra to assess the effect of security policies replacement and composition in OpenStack. Their solution can be considered as a proactive approach for checking operational property violations.

2.4   Taxonomy of Cloud Security Auditing

Based on the above-mentioned study on cloud security auditing, we devise a primary taxonomy for these works (as in Fig. 2.1). We consider the whole landscape from the perspective of their coverage and applied techniques. Therefore, we first categorize them based on their targeted cloud layers (e.g., data, user, virtual network, and SDN), then further identify various high-level security properties that these works support, and finally show their adopted approaches. Thus, it is trivial to understand which approaches are already explored for certain security problems under a particular cloud layer. Furthermore, our taxonomy can be useful towards building a fine-grained classification of cloud security auditing approaches.

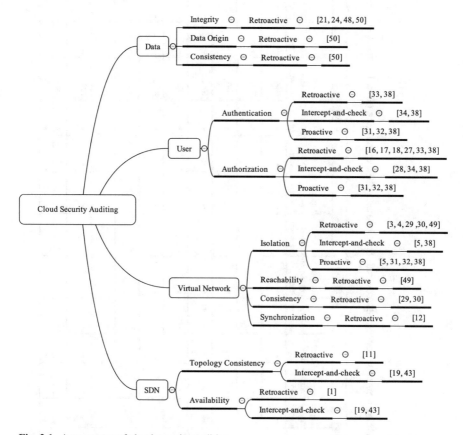

Fig. 2.1 A taxonomy of cloud security auditing

2.5   Comparative Study

This section shows a comparative study based on the taxonomy presented in the previous section. Table 2.1 summarizes the findings of this study. The first and second columns of the table listexisting works and their verification methods. The next four columns present their covered layers in the cloud. We mainly include works on four cloud layers: data, user, virtual network, and software-defined network (SDN). In the next three columns, we show the approaches (retroactive, intercept-and-check, and proactive) that a work adopts. Afterwards, there are five features enlisted to demonstrate the special skills of these works. The caching feature is marked when a work enables caching of verification results to enhance the efficiency of the auditing process. We mark the dependency model when a work utilizes the dependency relationship in the cloud to improve the efficiency and accuracy of the auditing process. The pre-computation step is to identify the works which perform a significant part of the verification step in advance

Table 2.1 Summary of existing cloud security auditing solutions highlighting their adopted methods, covered cloud layers, applied approaches, offered features, supported platforms and constraints

Proposals	Methods	Layers			Approaches				Features					Platforms			Adaptable to others	Constraints			
		Data	User	Virtual net.	SDN	Retroactive	Intercept-and-check	Proactive	Caching	Dependency model	Pre-computation	Batch auditing	Active auditing	Supporting OpenStack	Supporting Azure	Supporting VMware		After-the-fact	Prohibitive delay	Manual effort	Limited coverage
Wang et al. [116]	Cryptographic	•	–	–	–	•	–	–	–	–	–	•	–	–	–	–	•	•	N/A	–	–
Kai et al. [56]	Cryptographic	•	–	–	–	•	–	–	–	–	–	•	–	–	–	–	•	•	N/A	–	–
Doelit-zscher et al. [28]	Custom algorithm	–	–	•	–	•	–	–	–	–	–	–	–	•	–	–	•	•	N/A	–	–
Ullah et al. [112]	Custom algorithm	–	–	•	–	•	–	–	–	–	–	–	–	•	–	–	–	•	N/A	–	–
Solanas et al. [106]	Classifiers	–	•	–	–	•	–	–	–	–	–	–	–	•	–	–	–	•	N/A	–	–
Majumdar et al. [65]	CSP	–	•	–	–	•	–	–	–	–	–	–	–	•	–	–	–	•	N/A	–	–
Madi et al. [63, 64]	CSP	–	–	•	–	•	–	–	–	–	–	–	–	•	–	–	–	•	N/A	–	–
Cloud Radar [12]	Graph theory	–	–	•	–	•	–	–	–	–	–	–	–	•	–	•	–	•	N/A	–	–
Tenant-Guard [117]	Graph theory	–	–	•	–	•	–	–	–	–	–	–	–	•	–	–	–	•	N/A	–	–
SecGuru [9]	SMT	–	–	•	–	•	–	–	–	–	–	–	–	–	•	–	–	•	N/A	–	–
QRadar [48]	Custom	–	–	•	–	•	–	–	–	–	–	–	–	–	–	•	–	•	N/A	–	–

														N/A	
SPV [1]	Custom	–	–	–	–	–	–	–	•	–	–	•	–	–	–
Patron [62]	Custom algorithm	•	–	–	–	–	–	–	–	–	–	•	–	•	–
Weatherman (V1) [13]	Graph theory	–	•	•	•	–	–	–	–	–	–	•	•	•	•
Congress (V1) [89]	Datalog	•	•	•	–	–	•	–	–	•	–	•	•	•	–
TopoGuard [46, 105]	Custom	–	•	•	•	–	–	–	–	–	–	•	•	–	•
Majumdar et al. [69]	CSP + custom	•	–	•	–	–	–	–	–	–	–	•	•	–	–
Weatherman (V2) [13]	Graph theory	–	•	–	–	•	–	–	–	–	–	–	–	•	•
Congress (V2) [89]	Datalog	•	•	•	–	–	•	–	–	•	–	–	•	•	•
PVSC [66]	Custom algorithm	–	•	•	–	–	•	–	–	•	–	–	•	•	•
LeaPS [67]	Custom + Bayesian	•	•	•	–	–	•	–	–	•	–	–	–	•	•

The symbols (•), (–), and N/A mean supported/required, not supported/required, and not applicable, respectively. Note that, for both Weatherman and Congress, V1 and V2 refer to their proactive and intercept-and-check variants, respectively to reduce the response time of the runtime (usually in intercept-and-check and proactive) solutions. There exist few works which support auditing of multiple requests together. For them, we mark the batch auditing feature. The active auditing feature is an active-probing-based auditing solution which does not fully rely on the cloud provider for the audit data and instead actively participates in the targeted protocol to verify certain properties. The next four columns indicate the supporting cloud platforms for these auditing solutions. We mark the adaptable to others column when a work provides a detailed discussion on the process of porting the solution to different platforms. In the last four columns, we evaluate existing works based on the commonly observed constraints in the field of cloud security auditing. The after-the-fact constraint is marked if a work cannot prevent a security violation. The prohibitive delay is checked when a runtime work (i.e., intercept-and-check and proactive approaches) causes significant delay in responding to a user request. For retroactive solutions, we mark this column as not applicable (N/A). If a work involves significant manual effort (apart from the inputs from the users) in the auditing process, then we check the manual effort constraint. The limited coverage constraint is defined based on the expressiveness of an auditing solution. For instance, a work supporting first order logic to define security properties does not suffer from this constraint.

The key observations of this comparative study are as follows: First, there is no single auditing solution to verify multiple layers of the cloud. Therefore, today's cloud tenants require at least three different solutions to fulfill their auditing need, which might not be very usable for the tenants. Second, even though intercept-and-check approach is designed to prevent security violations, existing works under this category are not practical due to their prohibitive delay. Third, the proactive auditing approach is a promising solution to overcome the limitations of both retroactive and intercept-and-check approach. However, this approach still suffers from several practical issues, such as relying on manual efforts and limiting the expressiveness of security properties. Finally, there exist several features in the wild which significantly can improve the efficiency and accuracy of the auditing solution. However, there is a need of a unified solution with all these features at least to overcome major constraints. In the next chapters, we present solutions to the above-mentioned limitations.

Chapter 3
Auditing Security Compliance of the Virtualized Infrastructure

This chapter presents a security auditing approach for the cloud virtualized environment. More precisely, we focus primarily on virtual resources isolation based on structural properties (e.g., assignment of instances to physical hosts and the proper configuration of virtualization mechanisms), and consistency of the configurations in different layers of the cloud (infrastructure management layer, software-defined networking (SDN) controller layer, virtual layer, and physical layer). Although there already exist various efforts on cloud auditing (as shown in Chap. 2), to the best of our knowledge, none has facilitated automated auditing of structural settings of the virtual resources while taking into account the multi-layer aspects.

The security challenges faced by the cloud, mainly the loss of control and the difficulty to assess security compliance of the cloud service providers, leave potential customers still reluctant towards its adoption. These challenges stem from cloud-enabling technologies and characteristics. For instance, virtualization introduces complexity, which may lead to new vulnerabilities (e.g., incoherence between multiple management layers of hardware and virtual components). At the same time, concurrent and frequent updates needed to meet various requirements (e.g., workload balancing) may create even more opportunities for misconfiguration, security failures, and compliance compromises. Cloud elasticity mechanisms may cause virtual machines (VMs) belonging to different corporations and trust levels to interact with the same set of resources, causing potential security breaches [101]. Therefore, cloud customers take great interest in auditing the security of their cloud setup.

Security compliance auditing provides proofs with regard to the compliance of implemented controls with respect to standards as well as business and regulatory requirements. However, auditing in the cloud constitutes a real challenge. First, the coexistence of a large number of virtual resources on one side and the high frequency with which they are created, deleted, or reconfigured on the other side would require to audit, almost continuously, a sheer amount of information, growing continuously and exponentially [16]. Furthermore, a significant gap between the

© Springer Nature Switzerland AG 2019
S. Majumdar et al., *Cloud Security Auditing*, Advances in Information Security 76,
https://doi.org/10.1007/978-3-030-23128-6_3

high-level description of compliance recommendations (e.g., Cloud Control Matrix (CCM) [17] and ISO 27017 [52]) and the low-level raw logging information hinders auditing automation. More precisely, identifying the right data to retrieve from an ever increasing number of data sources and correctly correlating and filtering it constitute a real challenge in automating auditing in the cloud.

Motivating Example The following illustrates the challenges to fill the gap between the high-level description of compliance requirements as stated in the standards and the actual low-level raw audit data. In CCM [17], the control on Infrastructure and Virtualization Security Segmentation recommends *"isolation of business critical assets and/or sensitive user data, and sessions."* In ISO 27017 [52], the requirement on segregation in virtual computing environments mandates that *"cloud service customer's virtual environment should be protected from other customers and unauthorized users."* Moreover, the segregation in networks requirements recommends *"separation of multi-tenant cloud service customer environments."*

Clearly, any overlap between different tenants' resources may breach the above requirements. However, in an SDN/Cloud environment, verifying the compliance with standards and predefined policies requires gathering information from many sources at different layers of the cloud stack: the cloud infrastructure management system (e.g., OpenStack [90]), the SDN controller (e.g., OpenDaylight [82]), and the virtual components and verifying that effectively compliance holds in each layer. For instance, the logging information corresponding to the virtual network of tenant `0848cc1999-e542798` is available from at least these different sources:

- Neutron databases, e.g., records from table "Routers" associating tenants to their virtual routers and interfaces of the form `0848cc1999e542798` `(tenants_id)` ‖ `420fe1cd-db14-4780` (vRouter_id) ‖ `6d1f6103-9b7a-4789-ab16` (vInterface_id).
- Nova databases, e.g., records from table "Instances" associating VMs to their owners and their MAC addresses as follows: `0721a9ac-7aa1-4fa9` `(VM_ID)` ‖ `0848cc1999e542798` (tenants_id) and `fa:16:-3e:cd:b5:e1` `(MAC)` ‖ `0721a9ac-7aa1-4fa9` (VM_ID).
- Open vSwitch databases information, where ports and their associated tags can be fetched in this form `qvo4429c50c-9d` `(port_name)` ‖ `1084` (VLAN_ID).

As illustrated above, it is difficult to identify all the relevant data sources and to map information from those different sources at various layers to the standard's recommendations. Furthermore, potential inconsistencies in these layers make auditing tasks even more challenging. Additionally, as different sources may manipulate different identifiers for the same resource, correctly correlating all these data is critical to the success of the audit activity.

To facilitate automation, we present a compiled list of security properties relevant to the cloud virtualized environment that maps into different recommendations described in several security compliance standards in the field of cloud computing. Our auditing approach encompasses extracting configuration and logged information from different layers, correlating the large set of data from different origins, and finally relying on formal methods to verify the security properties and provide audit evidence. We furthermore implement the verification of these properties and show how the data can be collected and processed in the cloud environment with an application to OpenStack. Our approach shows scalability as it allows auditing a dataset of 300,000 virtual ports, 24,000 subnets, and 100,000 VMs in less than 8 s.

3.1 Auditing Approach for Virtualized Infrastructure

In this section, we present some preliminaries and describe our approach for auditing and compliance validation.

3.1.1 Threat Model

We assume that the cloud infrastructure management system has implementation flaws and vulnerabilities, which can be potentially exploited by malicious entities. For instance, a reported vulnerability in OpenStack Nova networking service, OSSN-0018/2014 [87], allows a malicious VM to reach the network services running on top of the hosting machine, which may lead to serious security issues. We trust cloud providers and administrators, but we assume that some cloud users and operators may be malicious [14]. We trust the cloud infrastructure management system for the integrity of the audit input data (e.g., logs, configurations, etc.) collected through API calls, events notifications, and database records (existing techniques on trusted auditing may be applied to establish a chain of trust from TPM chips embedded inside the cloud hardware to auditing components, e.g., [7]). We assume that not all tenants trust each other. They can either require not to share any physical resource with all the other tenants or provide a white (or black) list of trusted (or untrusted) customers that they are (not) willing to share resources with. Although our auditing framework may catch violations of specified security properties due to either misconfiguration or exploits of vulnerabilities, our focus is not on detecting specific attacks or intrusions.

3.1.2 Modeling the Virtualized Infrastructure

In a multi-tenant cloud infrastructure as a service (IaaS) model, the provider's physical and virtual resources are pooled to serve on demands from multiple customers. The IaaS cloud reference model [111] consists of two layers: The physical layer composed of networking, storage, and processing resources, and the virtualization layer that is running on top of the physical layer and enabling infrastructure resources sharing. Figure 3.1 refines the virtualization layer abstraction in [111] by considering tenant specific virtual resources such as virtual networks and VMs. Accordingly, a tenant can provision several VM instances and virtual networks. VMs may run on different hosts and be connected to many virtual networks through virtual ports. Virtualization techniques are used to ensure isolation among multiple tenants' boundaries. Host virtualization technologies enable running many virtual machines on top of the same host. Network virtualization mechanisms (e.g., VLAN and VXLAN) enable tenants' network traffic segregation, where virtual networking devices (e.g., Open vSwitches) play a vital role in connecting VM instances to their hosting machines and to virtual networks.

In addition to these virtual and physical resources illustrated as nodes, Fig. 3.1 shows the relationships between tenants' specific resources and cloud provider's resources. These relations will be used in Sect. 3.3 for the formalization of both the virtualized infrastructure model and the security properties. For instance, *IsAttachedOnPort* is a relationship with arity 3. It attaches a VM to a virtual subnet through a virtual port. This model can be refined with several levels of abstraction based on the properties to be checked.

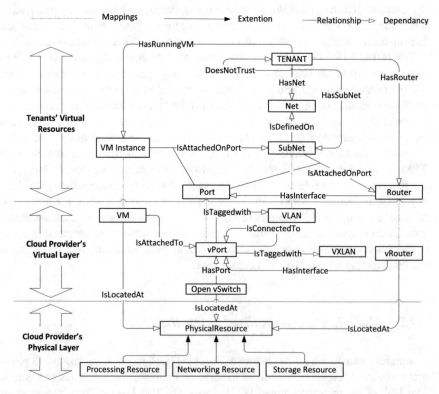

Fig. 3.1 A generic model of the virtualized infrastructures in the cloud

3.1.3   Cloud Auditing Properties

We classify virtualization-related properties into two categories: structural and operational properties. Structural properties are related to the static configuration of the virtualized infrastructure such as the assignment of instances to physical hosts, the assignment of virtual networking devices to tenants, and the proper configuration of isolation mechanisms such as VLAN configuration of each port. Operational properties are related to the forwarding network functionality. Those are mainly reachability-related properties such as loop-free forwarding and absence of black holes. Since the latter category has received significant attention in the literature (e.g., [10, 25, 120]), the former category constitutes the main focus of this chapter. As the major goal of this work is to establish a bridge between high-level guidelines in the security standards and low-level logs provided by current cloud systems, we start by extracting a list of concrete security properties from those standards and the literature in order to more clearly formulate the auditing problem. Table 3.1 presents an excerpt of the list of security properties we consider for auditing relevant standards (e.g., ISO 27002 [51], CCM [17]). Therein, we also classify properties based on their relevance to the stakeholders. In the following, we provide a brief description followed by an illustrating example for the sample properties, namely absence of common ownership of resources, no co-residence, and topology consistency.

Virtual Resource Isolation (No Common Ownership) Resource sharing technology was not designed to offer strong isolation properties for a multi-tenant architecture and thus has been ranked by the CSA among the nine notorious threats related to the cloud [2]. The related risks include the failure of logical isolation mechanisms to properly segregate virtual resources assigned to different tenants, which may lead to situations where one tenant has access to another tenant's resources or data. The no common ownership property aims at verifying that no virtual resource is co-owned by multiple tenants. Tenants are generally allowed to interconnect their own virtual resources to build their cloud virtual networks by modifying their configurations. However, if a virtual resource (e.g., a router or a port) is co-owned by multiple tenants, it can be part of several virtual networks belonging to different tenants, which can potentially create a breach of isolation.

Example 3.1 (No Common Ownership) This property has been violated in a real-life OpenStack deployment by exploiting the vulnerability OSSA-2014-008 [84] reported in the OpenStack Neutron networking service, which allows a tenant to create a virtual port on another tenant's router. An instance of our model can capture this violation as illustrated in Fig. 3.2. The model instance on the left side illustrates the initial entities and their relationships before exploiting the vulnerability. Assume that `Tenant_Beta`, by exploiting the said vulnerability, created `vPort_84`, and plugged it into `vRouter_A1`, which belongs to `Tenant_Alpha`. This would modify the model instance as illustrated on the right side showing the violation of *no common ownership*. Indeed, `Tenant_Beta` is the owner `vPort_84` as he

Table 3.1 An excerpt of security properties

Subject	Properties and sub-properties		Standards			
			ISO27002 [51]	ISO27017 [52]	NIST800 [77]	CCM [17]
Tenant	Data and proc. location correctness		18.1.1	18.1.1	IR-6, SI-5	SEF-01, IVS-04
	Virt. resource isolation (e.g., no common ownership)		–	–	–	STA-5
	Physical isolation (e.g., no co-residency)		–	13.1.3	SC-2	IVS-8, IVS-9
	Fault tolerance	Facility duplication	17.1, 17.2	12.1.3, 17.1, 17.2	PE-1, PE-13	BCR-03
		Storage service duplication				
		Redundant network connectivity				
Provider	No abuse of resources	Max number of VMs	–	–	–	IVS-11
		Max number of virtual networks				
	No resource exhaustion		–	–	–	IVS-05
Both	Topology consistency	Inf. management view/virtual inf.	–	13.1.3	SC-2	IVS-8, IVS-9
		SDN controller view/virtual inf.				

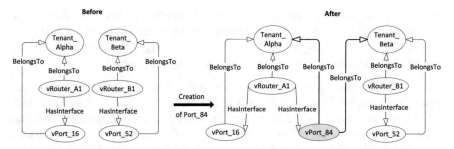

Fig. 3.2 Model instances for *no common ownership* property before and after the violation of *No Common Ownership* Property. After creating port vPort_84, the latter becomes owned by two tenants

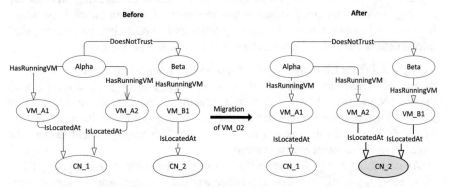

Fig. 3.3 Subsets of the virtual infrastructure model instances before and after violation of *no VM co-residency* property illustrating an example of data on VM locations. After migration, VM_A2 becomes co-resident with VM_B1 at compute node CN_2

is the initiator of the port creation. But since the port is connected to vRouter_A1, the created port would be considered as a common resource for both tenants.

Physical Isolation (No VM Co-residency) To maximize resources utilization, cloud providers consolidate virtual machines, possibly belonging to competing customers, to be run on the same physical machine, which may cause major security concerns as described in [122]. Physical isolation [44] aims at preventing side and covert channel attacks, and reducing the risk of attacks staged based on hypervisor and software switches vulnerabilities (e.g., [95]) by hosting VMs in different physical servers. Such attacks might lead to performance degradation, sensitive information leakage, and denial of service.

Example 3.2 (No VM Co-residency) Figure 3.3 consists of two subsets of instances of the virtual infrastructure model presented in Sect. 3.1.2. At the left side of the figure, we have two virtual machines VM_A1 and VM_A2 belonging to Tenant_Alpha and running at compute node CN_1, and VM_B1 owned by Tenant_Beta while running at compute node CN_2. Because of lack of trust, `Tenant_Alpha` may require physical isolation of his VMs from those of `Tenant_Beta`. However, as illustrated at the right side of Fig. 3.3, `VM_A2` can be migrated from `CN_1` to `CN_2` for load balancing. This new instance of the model after migration illustrates the violation of physical isolation.

Topology Consistency As stated in [49], it is critical to maintain consistency among cloud layers. The architectural model of the cloud can be described as a stack of layered services: physical layer, system resources layer, virtualized resources layer, support services layer, and at the top cloud-delivered services. Additionally, using SDN to implement network services increases management flexibility but also adds yet another layer in the stack. The presence of inconsistencies between these layers may lead to security breaches, which in turn makes the security controls at higher layers inefficient. Topology consistency consists of checking whether the topology view in the cloud infrastructure management system matches the actual implemented topology while considering different mappings between the physical infrastructure, the virtual infrastructure, and the tenants' boundaries.

Example 3.3 (Port Consistency) We suppose that a malicious insider managed to deliberately create a virtual port *vPort_40* on *Open_vSwitch_56* and label it with the VLAN identifier *VLAN_100* that is already assigned to tenant Alpha. This would allow the malicious insider to sniff tenant Alpha's traffic by mirroring the *VLAN_100* traffic to the created port, *vPort_40*. This clearly would lead to the violation of the network isolation property.

As illustrated in Fig. 3.4, we build two views of the virtualized topology: The actual topology is built based on data collected directly from the networking devices running at the virtualization layer (Open vSwitches), and the perceived topology is obtained from the infrastructure management layer (Nova and Neutron OpenStack databases). The dashed lines map one to one the entities between the two topologies (not all the mappings are shown for more readability). We can observe that *vPort_40* is attached to *VLAN_100*, which maps to *Net_01* (tenant Alpha's network), but there is no entity at the infrastructure management layer that maps to the entity *vPort_40* at the virtualization layer, which reveals a potential security breach.

Other Security Properties In the following, we briefly describe other security properties presented in Table 3.1.

- *Data and processing location correctness.* One of the main cloud-specific security issues is the increased complexity of compliance with laws and regulations [27]. The cloud provider might have data centers spread over different continents and governed by various court jurisdictions. Data and processing can be moved between the cloud provider's data centers without tenants' awareness, and fall under conflicting privacy protection laws.
- *Redundancy and fault tolerance.* Cloud providers have to apply several measures to achieve varying degrees of resiliency following the criticality of tenants' applications. Duplicating facilities in various locations and replicating storage services are examples of the measures that could be undertaken. Considering additional

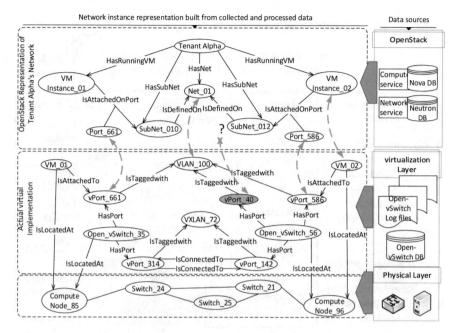

Fig. 3.4 Virtualized infrastructure model instance showing an OpenStack representation and the corresponding actual virtual layer implementation. VXLAN_72 and its ports are part of the infrastructure implementation and do not correspond to any component in tenant Alpha's resources

redundancy of network connectivity and information processing facilities has been mentioned in ISO 27002:2013 [51] as one of the best practices.

- *No abuse of resources.* Cloud services can be used by legitimate anonymous customers as a basis to illegitimately lead criminal and suspicious activities. For example, cloud services can be used to stage DDoS attacks [2].
- *No resource exhaustion.* The ease with which virtual resources can be provisioned in the cloud introduces the risk of resource exhaustion [76]. For example, creating a huge amount of VMs within a short time frame drastically increases the odds of misconfiguration, which opens up several security breaches [15].

3.2   Audit Ready Cloud Framework

Figure 3.5 illustrates a high-level architecture of our auditing framework. It has five main components: data collection and processing engine, compliance validation engine, audit report engine, dashboard, and audit repository database. The framework interacts mainly with the cloud management system, the cloud infrastructure system (e.g., OpenStack), and elements in the data center infrastructure to collect

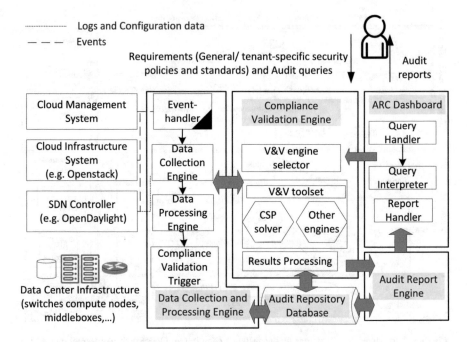

Fig. 3.5 A high-level architecture of our cloud auditing framework

various types of audit data. It also interacts with the cloud tenant to obtain the tenant requirements and to provide the tenant with the audit result. Tenant requirements encompass both general and tenant-specific security policies, applicable standards, as well as audit queries. For the lack of space, we will only focus on the following major components.

Our data collection and processing engine is composed of two sub-engines: the collection engine and the processing engine. The collection engine is responsible for collecting the required audit data in a batch mode, and it relies on the cloud management system to obtain the required data. The role of the processing engine is to filter, format, aggregate, and correlate this data. The required audit data may be distributed throughout the cloud and in different formats. The processing engine must pre-process the data in order to provide specific information needed to verify given properties. The last processing step is to generate the code for compliance validation and then store it in the audit repository database to be used by the compliance validation engine. The generated code depends on the selected back-end verification engine.

The compliance validation engine is responsible for performing the actual verification of the audited properties and the detection of violations, if any. Triggered by an audit request or updated inputs, the compliance validation engine invokes our back-end verification and validation algorithms. We use formal methods to capture formally the system model and the audit properties, which facilitates automated reasoning and is generally more practical and effective than manual inspection. If a security audit property fails, evidence can be obtained from the output of the verification back-end. Once the outcome of the compliance validation is ready, audit results and evidences are stored in the audit repository database and made accessible to the audit reporting engine. Several potential formal verification engines can serve our needs, and the actual choice may depend on the property being verified.

3.3 Formal Verification

As a back-end verification mechanism, we propose to formalize audit data and properties as Constraint Satisfaction Problems (CSP) and use a constraint solver, namely Sugar [108], to validate the compliance. CSP allows formulation of many complex problems in terms of variables defined over finite domains and constraints. Its generic goal is to find a vector of values (a.k.a. assignment) that satisfies all constraints expressed over the variables. If all constraints are satisfied, the solver returns SAT; otherwise, it returns UNSAT. In the case of a SAT result, a solution to the problem is provided. The key advantage of using CSP comes from the fact that it enables uniformly presenting the system's setup and specifying the properties in a clean formalism (e.g., first order logic (FOL) [8]), which allows to check a wide variety of properties [121]. Moreover using CSP avoids the state space traversal, which makes our approach more scalable for large datasets.

3.3.1 Model Formalization

Depending on the properties to be checked, we encode the involved instances of the virtualized infrastructure model as CSP variables with their domains definitions (over integer), where instances are values within the corresponding domain. For example, $Tenant$ is defined as a finite domain ranging over integer such that *(domain TENANT   0   max_tenant)* is a declaration of a domain of tenants, where the values are between 0 and *max_tenant*. Relations between classes and their instances are encoded as relation constraints and their supports, respectively. For example, $HasRunningVM$ is encoded as a relation, with a support as follows: (relation $HasRunningVM$ 2 $(supports(vm1, t1)(vm2, t2)))$. The support of this relation will be fetched and pre-processed in the data processing step. The CSP code mainly consists of four parts:

- *Variable and domain declaration.* We define different entities and their respective domains. For example, t is a variable defined over the domain $TENANT$, which range over integers.
- *Relation declaration.* We define relations over variables and provide their support from the audit data.

- *Constraint declaration*. We define the negation of each property in terms of predicates over the involved relations to obtain a counterexample in case of a violation.
- *Body*. We combine different predicates based on the properties to verify using Boolean operators.

3.3.2 Properties Formalization

Security properties would be expressed as predicates over relation constraints and other predicates. We express the sample properties in FOL. Table 3.2 summarizes the predicates required for expressing the properties. Those predicates correspond to CSP relation constraints used to describe the current configuration of the system. Note that predicates that do not appear as relationships in Fig. 3.1 are inferred by correlating other available relations.

No Common Ownership We check that a tenant specific virtual resource belongs to a unique tenant.

$$\forall r \in \text{Resource}, \forall t1, t2 \in \text{TENANT} \quad (3.1)$$

$$\text{BelongsTo}(r, t1) \wedge \text{BelongsTo}(r, t2) \quad \rightarrow (t1 = t2)$$

No Co-residence Based on the collected data, we check that the tenant's instances are not co-located in the same compute node with adversaries' instances.

$$\forall t1, t2 \in \text{TENANT}, \forall vm1, vm2 \in \text{INSTANCE}, \quad (3.2)$$

Table 3.2 First order logic predicates

Relations in properties	Evaluate to *True* if
BelongsTo(r, t)	The resource r is owned by tenant t
HasRunningVM(vm, t)	The tenant t has a running virtual machine vm
DoesNotTrust(t1, t2)	Tenant $t2$ is not trusted by tenant $t1$ which means that $t1$'resources should not share the same hardware with $t2$' instances
IsLocatedAt(vm, cn)	The instance vm is located at the compute node cn
IsAssignedPortVLAN (p,v,t)	The port p is assigned to the VLAN v which is in turn assigned to tenant t
HasPortVLAN(vs, p, v)	The port p is created at the virtual switch vs and assigned to VLAN v

$$\forall cn1, cn2 \in \text{COMPUTEN} :$$

$$\text{HasRunningVM}(vm1, t1) \wedge \text{HasRunningVM}(vm2, t2) \wedge$$

$$\text{DoesNotTrust}(t1, t2) \wedge \text{IsLocatedAt}(vm1, cn1) \wedge$$

$$\text{IsLocatedAt}(vm2, cn2) \rightarrow cn1 \neq cn2$$

Topology Consistency We check that mappings between virtual resources over different layers are properly maintained and that the current view of the cloud infrastructure management system on the topology matches the actual topology of the virtual layer. In the following, we consider port consistency as a specific case of topology consistency. We check that the set of virtual ports assigned to a given tenant's VLAN by the provider correspond exactly to the set of ports inferred from data collected from the actual infrastructure's configuration for the same tenant's VLAN.

$$\forall vs \in \text{vSWITCH}, \quad \forall p \in \text{Port} \quad \forall t \in \text{TENANT} \quad \forall v \in \text{VLAN} \quad (3.3)$$

$$\text{HasPortVlan}(vs, p, v) \Leftrightarrow \text{IsAssignedPortVLAN}(p, v, t)$$

Example 3.4 Listing 3.1 is the CSP code to verify the no common ownership, no co-residence, and port consistency properties for our running example. Variables along with their respective domains are first declared (see Listing 3.1 lines 2–10). Based on the properties of interest, a set of relations are defined and populated with their supporting tuples, where the support is generated from actual data in the cloud (see lines 12–17). Then, the properties are declared as predicates over these relations (see lines 19–21). Finally, the disjunction of the predicates is instantiated for verification (see lines 22–23). As we are formalizing the negation of the properties, we are expecting the UNSAT result, which means that none of the properties holds (i.e., no violation of the properties). We present the verification outputs in Sect. 3.4.

Listing 3.1 Sugar source code for common ownership, co-residence, and port consistency verification

```
1    // Declaration
2    (domain TENANT 0 60000) (domain RESOURCE 0 216000)
3    (domain INSTANCE 0 100000) (domain HOST 0 1000)
4    (domain PORT 0 300; 000) (domain VLAN 0 60000)
5    (domain VSWITCH 0 1000)
6    ( int  T1 TENANT) (int T2 TENANT)
7    ( int  R1 Resource) ( int  R2 Resource)
8    ( int  VM1 INSTANCE) (int VM2 INSTANCE)
9    ( int  H1 HOST) (int H2 HOST)(int V VLAN)
10   ( int  T TENANT) (int P PORT) (int vs  VSWITCH)
11   // Relations  Declarations  and Audit data  as  their  support
```

₁₂ (relation BelongsTo 2 (supports (18037 10)(18038 10)(18039 10)   (18040
          10)(18038 11)(18042 11)(18043 11)(18044 11)(18045 11)(18046 12)(18047
          12)))
₁₃ (relation HasRunningVM 2 (supports (6100 10)(6101 10)(6102 11) (6103
          11)(6104 11)(6105 11)))
₁₄ (relation IsLocatedAt 2 (supports (((6089 11000)(6090 11000)(6093 11000)(6101
          11100)(6102 11100))
₁₅ (relation DoesNotTrust 2 (supports (9 11)(9 13)(9 14)))
₁₆ (relation IsAssignedPortVLAN 3 (supports (18028 6017 9)(18029 6018 9) (18030
          6019 10)(18031 6019 10)(18032 6020 10)))
₁₇ (relation HasPortVLAN 3 (supports(511 18030 6019)(511 18031 6019 10) (512
          18032 6020)(512 18033 6021)))
₁₈ // Security   properties expressed in terms of predicates over relations
₁₉ (predicate (CommonOwnership T1 R1 T2 R2)(and (BelongsTo T1 R1) (BelongsTo
          T2 R2) (= R1 R2) (not (= T1 T2))))
₂₀ (predicate (coResidence T1 T2 VM1 VM2 H1 H2) (and (DoesNotTrust T1 T2)
          (HasRunningVM VM1 T1)(HasRunningVM VM2 T2) (IsLocatedAt H1 VM1)
          (IsLocatedAt H2 VM2) (=H1 H2)))
₂₁ (predicate (portConsistency P V T)(or (and (IsAssignedPoprtVLAN P V T)
          (not(HasPortVLAN VS P V))) (and (HasPortVLAN VS P V)
          (not(IsAssignedPoprtVLAN P V T)))))
₂₂ // The Body
₂₃ (or (CommonOwnership T1 R1 T2 R2) (coResidence T1 T2 VM1 VM2 H1 H2)
          (portConsistency P V T))

3.4 Application to OpenStack

This section describes how we integrate our audit and compliance framework into
OpenStack. First, we briefly present the OpenStack networking service (Neutron),
the compute service (Nova), and Open vSwitch [114], the most popular virtual
switch implementation. We then detail our auditing framework implementation and
its integration in OpenStack along with the challenges that we faced and overcame.

3.4.1 Background

OpenStack [90] is an open-source cloud infrastructure management platform that
is being used almost in half of private clouds and significant portions of the public
clouds (see [22] for detailed statistics). The major components of OpenStack to
control large collections of computing, storage, and networking resources are,
respectively, Nova, Swift, and Neutron along with Keystone. Following is the brief
description of Nova and Neutron:

Nova [90] This is the OpenStack project designed to provide massively scalable, on demand, self-service access to compute resources. It is considered as the main part of an infrastructure as a service model.

Neutron [90] This OpenStack system provides tenants with capabilities to build rich networking topologies through the exposed API, relying on three object abstractions, namely networks, subnets, and routers. When leveraged with the modular layer 2 plug-in (ML2), Neutron enables supporting various layer 2 networking technologies. For our testbed we consider Open vSwitch as a network access mechanism and we maintain two types of network segments, namely VLAN for communication inside of the same compute node and VXLAN for inter-compute nodes communications.

Open vSwitch [114] Open vSwitch is an open-source software switch designed to be used as a vSwitch in virtualized server environments. It forwards traffic between different virtual machines (VMs) on the same physical host and also forwards traffic between VMs and the physical network.

3.4.2   Integration to OpenStack

We focus mainly on three components in our implementation: the data collection engine, the data processing engine, and the compliance validation engine. The data collection engine involves several components of OpenStack, e.g., Nova and Neutron for collecting audit data from databases and log files, different policy files and configuration files from the OpenStack ecosystem, and log files from various virtual networking components such as Open vSwitch to fully capture the configuration. The data is then converted into a consistent format and missing correlation is reconstructed. The results are used to generate the code for the validation engine based on Sugar input language. The compliance validation engine performs the verification of the properties by feeding the generated code to Sugar. Finally, Sugar provides the results on whether the properties hold or not. Figure 3.6 illustrates the steps of our auditing process. In the following, we describe our implementation details along with the related challenges.

Data Collection Engine We present hereafter different sources of data in OpenStack along with the current support for auditing offered by OpenStack and the virtual networking components. The main sources of audit data in OpenStack are logs, configuration files, and databases. Table 3.3 shows some sample data sources. The involved sources for auditing depend on the objective of the auditing task and the tackled properties. We use three different sources to audit configuration correctness of virtualized infrastructures:

- *OpenStack.* We rely on a collection of OpenStack databases, hosted in a MySQL server, that can be read using component-specific APIs such as Neutron APIs. For instance, in Nova database, table *Compute-node* contains information about

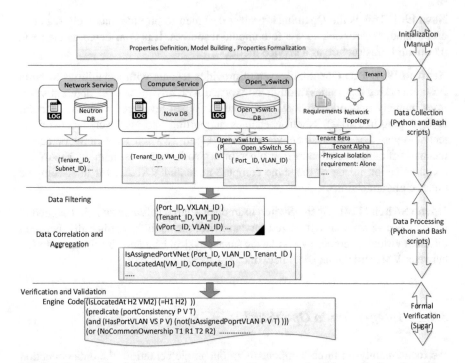

Fig. 3.6 Our OpenStack-based auditing solution with the example of data collection, formatting, correlation building, and Sugar source generation

Table 3.3 Sample data sources in OpenStack, Open vSwitch, and tenants' requirements

Relations	Sources of data
BelongsTo	Table *Instances* in Nova database and *Routers, Subnets* and *Ports* in Neutron database, Neutron logs
DoesnotTrust	The tenant physical isolation requirement input
IsLocatedAt	Tables *Instances* in Nova database
IsAssignedPortVLAN	*Networks* in Nova database and *Ports* in Neutron database
HasPortVLAN	Open vSwitch instances located at various compute nodes
HasRunningVM	Table *Instances* in Nova database

the hosting machines such as the hypervisor's type and version, table *Instance* contains information about the project (tenant) and the hosting machine, table *Migration* contains migration events' related information such as the source compute and the destination compute. The Neutron database includes various information such as security groups and port mappings for different virtualization mechanisms.

- *Open vSwitch*. Flow tables and databases of Open vSwitch instances located in different compute nodes and in the controller node constitute another important source of audit data for checking whether there exist any discrepancies between the actual configuration and the OpenStack view.

- *Tenant policies*. We consider security policies expressed by the customers, such as physical isolation requirements. As expressing tenants' policies is out of the scope of this solution, we assume that they are parsable XML files.

Data Processing Engine Our data processing engine, which is implemented in Python, mainly retrieves necessary information from the collected data according to the targeted properties, recovers correlation from various sources, eliminates redundancies, converts it into appropriate formats, and finally generates the source code for Sugar.

- Firstly, for each property, our plug-in identifies the involved relations. The relations' support is either fetched directly from the collected data such as the support of the relation *BelongsTo* or recovered after correlation, as in the case of the relation *IsAssignedPortVLAN*.
- Secondly, our processing plug-in formats each group of data as an n-tuple, i.e., *(resource, tenant),(port, vlan, tenant)*, etc.
- Finally, our plug-in uses the n-tuples to generate the portions of Sugar's source code, and append the code with the variable declarations, relationships, and predicates for each security property (as discussed in Sect. 3.3). Different scripts are needed to generate Sugar source code for the verification of different properties.

Compliance Validation The compliance validation engine is discussed in detail in Sect. 3.3. In the following example, we discuss how our auditing framework can detect the violation of the no common ownership, no co-residence, and port inconsistency security properties caused by the attack scenarios of our running example.

Example 3.5 In this example, we describe how a violation of no common ownership, no co-residence, and port consistency properties may be caught by auditing.

Firstly, our program collects data from different tables in the Nova and Neutron databases, and logs from different Open vSwitch instances. Then, the processing engine correlates and converts the collected data and represents it as tuples; for an example: (18038 10) (6100 11000) (512 6020 18033), where Port_84: 18038, Alpha: 10, VM_01: 6100, Open_vSwitch_56: 512, vPort_40: 18033, and VLAN_100: 6020. Additionally, the processing engine interprets each property and generates the associated Sugar source code (see Listing 3.1 for an excerpt of the code) using processed data and translated properties. Finally, Sugar is used to verify the security properties.

We show for each property how the violation is detected:

- No common Ownership. The predicate *CommonOwnership* will evaluate to true if there exists a resource belonging to two different tenants. As Port_84 has been created by Beta, *BelongsTo(Port_84, Beta)* evaluates to true based on collected data from Neutron logs. Port_84 is defined on Alpha's router;

hence, *BelongsTo(Port_84, Alpha)* evaluates to true based on collected data from Neutron database. Consequently, the predicate *CommonOwnership* evaluates to true. In this case, the output of sugar (SAT) is the solution of the problem (r1 = 18038; r2 =18038; t1 =10; t2=11), which is actually the proof that Port_84 violates the no common ownership property.

- No co-residence. In our example (see Fig. 3.3), the supports *HasRunningVM((VM_02, Alpha)(VM_03, Beta))*, *IsLocatedAt((VM_02, Compute_Node_96)(VM_03,Compute_Node_96)*, and *DoesNotTrust(Alpha, Beta)*, where VM_02:6101, VM_03:6102, and Compute_Node_96:11100, make the predicate evaluate to true meaning that the no co-residence property has been violated.

- Port consistency. The predicate *PortConsistency* evaluates to true if there exists a discrepancy between the OpenStack view of the virtualized infrastructure and the actual configuration. The support HasPortVLAN(Open_vSwitch_56, vPort_40, VLAN_100) makes the predicate evaluate to true, as long as there is no tuple such that *IsAssignedPortVLAN (Port, VLAN_100, Alpha)* where *Port* maps to vPort_40:18033.

Challenges Checking the configuration correctness in virtualized environment requires considering logs generated by virtualization technologies at various levels, and checking that mappings are properly maintained over different layers. Unfortunately, OpenStack does not maintain such overlay details.

At the OpenStack level, ports are directly mapped to VXLAN IDs, whereas at the Open-vSwitch level, ports are mapped to VLAN tags and mappings between the VLAN tags and VXLAN IDs are maintained. To overcome this limit, we devised a script that generates logs from all the Open vSwitch instances. The script recovers mappings between VLAN tags and the VXLAN IDs from the flow tables using the *ovs-ofctl* command line tool. Then, it recovers mappings between ports and VLAN tags from the Open-vSwitch database using the *ovs-vsctl* command line utility.

Checking the correct configuration of overlay networks requires correlating information collected both from Open vSwitch instances running on top of various compute nodes and the controller node, and data recovered from OpenStack databases. To this end, we extended our data processing plug-in to deduce correlation between data. For example, we infer the relation (*por t vlan tenant*) from the available relations (*vlan vxlan*) recovered from Open vSwitch and (*port vxlan tenant*) recovered from the Nova and Neutron databases. In our settings, we consider a ratio of 30 ports per tenant, which leads to 300,000 entries in the relation (*port vxlan tenant*) for 10,000 tenants. The number of entries is considerably larger than the number of tenants, because a tenant may have several ports and virtual networks. As a consequence, with the increasing number of tenants, the size of this relation grows and complexity of the correlation step also increases proportionally. Note that, correlation is required for several of our listed properties.

An auditing solution becomes less effective if all needed audit evidences are not collected properly. Therefore, to be comprehensive in our data collection process, we firstly check fields of all varieties of log files available in OpenStack, all configuration files, and all Nova and Neutron database tables. Through this process, we identify all possible types of data with their sources.

3.5 Experiments

In this section, we discuss the performance of our work by measuring the execution time, memory, and CPU consumption.

3.5.1 Experimental Setting

We deployed OpenStack with one controller node and three compute nodes, each having Intel i7 dual core CPU and 2 GB memory running Ubuntu 14.04 server. Our OpenStack version is DevStack Juno (2014.2.2.dev3). We set up a real testbed environment constituted of 10 tenants, 150 VMs, and 17 routers. To stress the verification engine and assess the scalability of our approach as a whole, we furthermore simulated an environment with 10,000 tenants, 100,000 VMs, 40,000 subnets, 20,000 routers, and 300,000 ports with a ratio of 10 VMs, 4 subnets, 2 routers, and 30 ports per tenant. To verify compliance, we use the V&V tool, Sugar V2.2.1 [108]. We conduct the experiment for 20 different audit trail datasets in total. All data processing and V&V experiments are conducted on a PC with 3.40 GHz Intel Core i7 Quad core CPU and 16 GB memory and we repeat each experiment 1000 times.

3.5.2 Results

The first set of our experiment (see Fig. 3.7) demonstrates the time efficiency of our auditing solution. Figure 3.7a illustrates the time in milliseconds required for data processing and compliance verification steps for port consistency, no co-residence, and no common ownership properties. For each of the properties, we vary the most significant parameter (e.g., the number of ports, VMs, and subnets for port consistency, no co-residence, and no common ownership properties, respectively) to assess the scalability of our auditing solution. Figure 3.7b (left) shows the size of the collected data in KB for auditing by varying the number of tenants. The collected data size reaches to around 17 MB for our largest dataset. We also measure the time for collecting data as approximately 8 min for a fairly large cloud setup. (10,000 tenants, 100,000 VMs, 300,000 ports, etc.) Note that data collection time heavily depends on deployment options and complexity of the setup. Moreover, the initial data collection step is performed only once for the auditing process (later on incremental collection will be performed at regular intervals), so the time may be

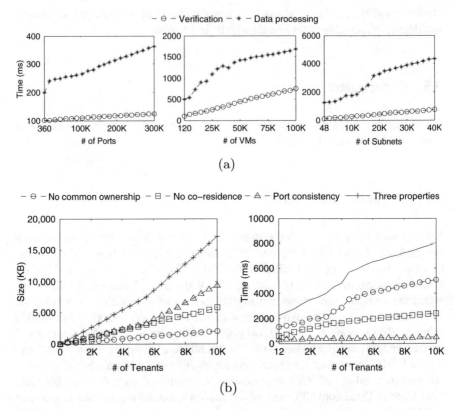

Fig. 3.7 Execution time for each auditing step, total size of the collected audit data, and total time for different properties using our framework. (**a**) Time required for data processing and verification for the port consistency (left), no co-residence (middle), and no common ownership (right) by varying number of ports, VMs, and subnets, respectively. (**b**) Total size (left) of collected audit data and time required (right) for auditing the port consistency, no co-residence, no common ownership and sequentially auditing three properties (worst case) by varying number of tenants

considered reasonable. Figure 3.7b (right) shows the total execution time required for each property individually and in total. Auditing no common ownership property requires the longest time, because of the highest number of predicates used in the verification step; however, it finishes in less than 4 s. In total, the auditing of three properties completes within 8 s for the largest dataset, when properties are audited sequentially. However, since there is no interdependency between verifying different security properties, we can easily run parallel verification executions. The parallel execution of the verification step for different properties reduces the execution time to 4 s, the maximum verification time required among three security properties. Additionally, we can infer that the execution time is not a linear function of the number of security properties to be verified. Indeed, auditing more security properties would not lead to a significant increase in the execution time.

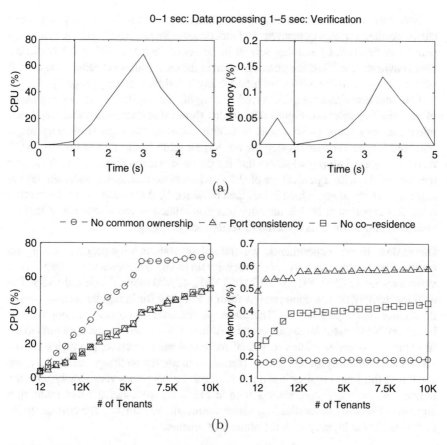

Fig. 3.8 CPU and memory usage for each step and for different properties of our auditing solution over time. (**a**) Peak CPU and memory usage for each step of our auditing solution over time when there are 10,000 tenants, 40,000 subnets, 100,000 VMs, and 300,000 ports. (**b**) CPU (left) and memory (right) usage for each step of our auditing solution over time when there are 6000 tenants, 24,000 subnets, 60,000 VMs, and 180,000 ports

The objective of our second experiment (Fig. 3.8a (left), b (left)) is to measure the CPU usage (in %). In Fig. 3.8a (left), we measure the peak CPU usage consumed by data processing and verification steps while auditing the no common ownership property. We notice that the average CPU usage is around 35% for the verification, whereas it is fairly negligible for the data processing step. According to Fig. 3.8b (left), the CPU usage grows almost linearly with the number of tenants. However, the speed of increase varies depending on the property. It reaches a peak of over 70% for the no common ownership property for 10,000 tenants. This is due to the huge amount of tenant specific resources (e.g., for 10,000 tenants the number of the resources involved may reach 216,000).

Note that, we conduct our experiments in a single PC; if the security properties can be verified through concurrent and independent Sugar executions, we can easily parallelize this task by running several instances of Sugar on different VMs in the cloud environment. Thus the parallelization in the cloud allows to reduce the overall verification time to the maximum time for any individual security property.

Our final experiment (Fig. 3.8a (right), b (right)) demonstrates the memory usage of our auditing solution. Figure 3.8a (right) shows that data processing step has a minor memory usage (with a peak of 0.05%), whereas the highest memory usage observed for the verification step for our largest setup is less than 0.19% of 16 GB memory. Figure 3.8b (right) shows that the port consistency property has the lowest memory usage with a percentage of 0.2%, whereas no common ownership has the highest memory usage, which is less than 0.6% for 10,000 tenants. Our observation from this experiment is that memory usage is related to the number of relations, variables, and constraints involved to verify each property.

Discussion In our experiments, we audited several security properties, e.g., no common ownership and port consistency, for up to 10,000 tenants with a large set of various resources (300,000 ports, 100,000 VMs, 40,000 subnets) in less than 8 s. The auditing activity occurs upon request from the auditor (or in regular intervals when the auditor sets regular audits). Therefore, we consider the costs of our approach to be reasonable even for large data centers. Although we report results for a limited set of security properties related to virtualized cloud infrastructure, promising results show the potentiality of the use of formal methods for auditing. Particularly, we show that the time required for our auditing solution grows very slowly with the number of security properties. As seen in Fig. 3.7a, we anticipate that auditing a large list of security properties in practice would still be realistic. The cost generally increases almost linearly with the number of tenants.

3.6   Conclusion

In this chapter, we elaborated a generic model for virtualized infrastructures in the cloud. We identified a set of relevant structural security properties to audit and mapped them to different standards. Then, we presented a formal approach for auditing cloud virtualized infrastructures from the structural point of view. Particularly, we showed that our approach is able to detect topology inconsistencies that may occur between multiple control layers in the cloud. Our evaluation results show that formal methods can be successfully applied for large data centers with a reasonable overhead. However, this work does not take into account the complexity factor and multi-layered nature of the cloud, and handle the cross-layer consistent isolation verification. In the next chapter, we overcome this limitation through a dedicated auditing framework for cross-layer virtual network isolation verification.

Chapter 4
Auditing Virtual Network Isolation Across Cloud Layers

In this chapter, taking into account the complexity factor and multi-layered nature of the cloud, we present an automated cross-layer approach that tackles the above issues for auditing isolation requirements between virtual networks in a multi-tenant cloud. We focus on isolation at layer 2 virtual networks and overlay, namely topology isolation, which is the basic building block for networks communication and segregation for upper network layers.[1] To the best of our knowledge, this is the first effort on auditing cloud infrastructure isolation at layer 2 virtual networks and overlay taking into account cross-layer consistency in the cloud stack.

Security and privacy concerns are still holding back the widespread adoption of the cloud [100]. Particularly, multi-tenancy in cloud environments, supported by virtualization, allows optimal and cost-effective resource sharing among tenants that do not necessarily trust each other. Furthermore, the highly dynamic, elastic, and self-service nature of the cloud introduces additional operational complexity that may cause several misconfigurations and vulnerabilities, leading to violations of baseline security and non-compliance with security standards (e.g., ISO 27002/27017 [51, 52] and CCM 3.0.1 [17]). Particularly, network isolation failures are among the foremost security concerns in the cloud [18, 23]. For instance, virtual machines (VMs) belonging to different corporations and trust levels may share the same set of resources, which opens up opportunities for inter-tenant isolation breaches [101]. Consequently, cloud tenants may raise questions like: "How to make sure that all my virtual resources and private networks are properly isolated from other tenants' networks, especially my competitors? Are my vertical segment (e.g., for finance, human resources, etc.) properly segregated from each other?".

Security auditing aims at verifying that the implemented mechanisms are actually providing the expected security features. However, auditing security without suitable automated tools could be practically infeasible due to the design complexity

---

[1] We refer to the network layers defined in the Open Systems Interconnection (OSI) model.

© Springer Nature Switzerland AG 2019
S. Majumdar et al., *Cloud Security Auditing*, Advances in Information Security 76,
https://doi.org/10.1007/978-3-030-23128-6_4

and the sheer size of the cloud as motivated in the following example. Note that domain-specific terms used in this chapter are summarized in the glossary.

Motivating Example Figure 4.1 illustrates a simplified view of an OpenStack [90] configuration example for virtualized multi-tenant cloud environments. Following a layered architecture [98], the cloud stack includes an infrastructure management layer responsible of provisioning, interconnecting, and decommissioning a set of virtual resources belonging to different tenants, at the implementation layer, across distributed physical resources. For instance, at the infrastructure management layer, virtual machines `VM_Adb` and `VM_Bapp1` are defined in separate virtual networks, `vNet_A` and `vNet_B` belonging to `Tenant_Alpha` and `Tenant_Beta`, respectively. At the implementation layer, these VMs are instantiated on `Physical Server_1` as `VM_11` and `VM_21` and are interconnected to form those virtual networks. As the latter networks share the same physical substrate, network isolation mechanisms are defined at the management layer and configured at the implementation layer through network virtualization mechanisms to ensure their logical segregation. For instance, virtual local area network (VLAN) is used to isolate different virtual networks at the host level (more details are provided in Sect. 4.1.1). To audit isolation as defined in applicable standards, there exist several challenges.

- The gap between the high-level description of the requirements in the standards and the actual security properties hinders auditing automation. For instance, the requirement on segregation in networks in ISO 27017 [52] recommends *"separation of multi-tenant cloud service customer environments"*. Stated as such, these requirements do not detail exactly what data to be checked or how it should be verified.

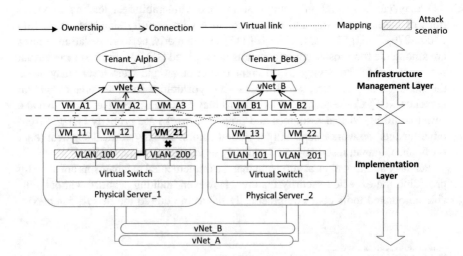

Fig. 4.1 A two-layer view of a multi-tenant virtualized infrastructure in cloud: the infrastructure management layer and the implementation layer

- The layered nature of the cloud stack and the dependencies between layers make existing approaches that separately verify each single layer ineffective. Those layers maintain different but complementary views of the virtual infrastructure and current isolation mechanisms configurations. For instance, assume Tenant_Beta compromises the hypervisor on Physical Server_1 (e.g., by exploiting some vulnerabilities [95]) and succeeds to directly modify VLAN_200 associated with VM_21 to become VLAN_100 that is currently associated with VM_11 and VM_12 on Physical Server_1. This leads to a topology isolation breach as both VMs will become part of the same layer 2 virtual network defined for vNet_A, opening the door for further attacks [104]. The verification of the management layer view cannot detect such a breach as VLAN tags are managed locally at the implementation layer. Additionally, verifying the implementation layer only without mapping the virtual resources to their owners (maintained only at the management layer) would not allow a per-tenant identification of the breached resource. For example, the association between VM_Bapp1, vNet_B, and their owner (Tenant_Beta) in the management layer view should be consistently mapped into the association between VM_21 in Physical Server_1 with VLAN_200 at the implementation level. This should be done for all tenants. Considering the implementation layer after the attack in Fig. 4.1, VM_11, VM_12, and VM_21 in Physical Server_1 can be identified to be on the same VLAN, namely VLAN_100. However, without considering that the corresponding VMs at the management layer are in different virtual networks and belong to different tenants, the breach cannot be properly detected.
- Correctly identifying the relevant data and their sources in the cloud for each security requirement increases the complexity of auditing. This can be amplified with the diversity and plurality of data sources located at different cloud-stack layers. Furthermore, the data should not be collected only from different layers but also from different physical servers. In addition, their underlying semantics and relationships should be properly understood to be able to process it. The relation of this data and its semantics to the verified property constitutes a real challenge in automating cloud auditing.

In the rest of this chapter, we discuss how these challenges can be addressed. First, to fill the gap between standards and isolation verification, we devise a set of concrete security properties based on the literature and common knowledge on layer 2 virtual networks isolation and relate them to relevant requirements in security standards. Second to identify the relevant data for auditing network isolation and capture its underlying semantics across multiple layers, we elaborate a model capturing the cloud-stack layers and the verified network layers along with their inter-dependencies and isolation mechanisms. To the best of our knowledge, we are the first to propose such a model. Third, we present an off-line verification approach that spans the OpenStack implementation and management layers, which allows to evaluate the consistency of layer 2 virtual network isolation. We rely on the model defined above as input to our approach and a constraint satisfaction problem (CSP) solver, namely Sugar [108], as a back-end verification tool. Fourth, we report real-life experience and challenges faced when integrating our auditing and compliance validation solution into OpenStack. Finally, we conduct experiments to demonstrate the applicability of our approach.

4.1 Models

In this section, we provide a background on the network isolation mechanisms considered in this chapter, and we present the threat model followed by our model that captures tenants' virtual networks at the infrastructure management and implementation layers.

4.1.1 Preliminaries

In this work, we focus on layer 2 virtual networks deployed in cloud environments managed by OpenStack. We furthermore consider Open vSwitch (OVS)[2] for providing layer 2 network function to guest VMs at the host level [97].

In large scale OpenStack-based cloud infrastructures, layer 2 virtual networks are implemented on the same server using virtual LANs (VLAN), and across the physical network through virtual extended LAN (VXLAN) as an overlay technology. The VXLAN technology is used to overcome the scale limitation of VLANs, which only allows for a maximum of 4096 tags [23]. More specifically, on each physical server, disjoint VLAN tags are assigned to ports connecting VMs that are part of different isolated virtual networks. Furthermore, a unique VXLAN identifier is assigned per isolated virtual network in order to extend layer 2 virtual networks between different physical servers, thus forming an overlay network. When the traffic leaves a VM (or a physical server), the appropriate VLAN tag (or VXLAN identifier) is inserted into the traffic by configurable OVS forwarding rules to maintain proper layer 2 traffic isolation. The mapping between VLAN tags and VXLAN identifiers performed by the OVS rules ensures that the traffic is smoothly steered between sources and destinations deployed over different physical servers.

Example 4.1 Figure 4.2 illustrates a more detailed view of layer 2 virtual networks implementation for the configuration showed in Fig. 4.1. According to the latter figure, VM_11, VM_12, and VM_13 belong to Tenant_Alpha and are connected to vNet_A. VLAN_100 is defined at Physical Server_1 to enable isolated layer 2 communication between VM_11 and VM_12, whereas VLAN_200 is defined to isolate VM_21 at the same physical server since the latter

---

[2]Open vSwitch OVS is one of the mostly used OpenFlow-enabled virtual switches in more than 30% deployments, and is compatible with most hypervisors including Xen, KVM, and VMware.

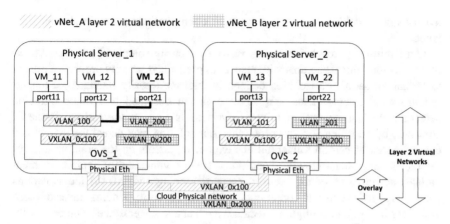

Fig. 4.2 A detailed view of the implementation layer of Fig. 4.1

VM is connected to another virtual network (vNet_B). Similarly, at Physical Server_2, different VLAN tags, namely VLAN_101 and VLAN_201, are defined to isolate VM_13 and VM_22, respectively, since they are connected to different networks. Since VM_11, VM_12, and VM_13 are all connected to the same virtual network (see Fig. 4.1) but deployed over two different physical servers, VXLAN is used as an overlay protocol to logically connect VMs across physical servers while ensuring isolation. To this end, two distinct VXLAN identifiers, namely VXLAN_0×100 and VXLAN_0×200, are associated to vNet_A and vNet_B, respectively. Then, to achieve end-to-end isolation, VXLAN_0×100 is attached to VLAN_100 on Physical Server_1 and to VLAN_101 on Physical Server_2, while VXLAN_0×200 is attached to VLAN_200 on Physical Server_1 and to VLAN_201 on Physical Server_2. This would allow to isolate the virtual networks both at the host level (through different VLAN tags) and at the physical network level (through different VXLAN identifiers).

4.1.2 Threat Model

We assume that the cloud infrastructure management system has implementation flaws and vulnerabilities, which can be potentially exploited by malicious entities leading to tenants' virtual infrastructures isolation failures. For instance, a reported vulnerability in OpenStack Neutron OSSA-2014-008 [85] allows a tenant to create a virtual port on another tenant's virtual router without checking his identity. Exploiting such vulnerabilities leads to serious isolation breaches opening doors to more harmful attacks such as network sniffing. As another example, a malicious tenant can take advantage from the known cloud data centers configuration strategies to locate his victim inside the cloud [101]. In addition, he can compromise some host hypervisors to deliberately change network configurations at the implementation layer.

Our auditing approach focuses on verifying security compliance of OpenStack-managed cloud infrastructures with respect to predefined security properties related to virtual infrastructure isolation defined in relevant security standards or tenant-specific requirements. Thus, our solution is not designed to replace intrusion detection systems or vulnerability analysis tools (e.g., vulnerability scanners). However, by verifying security properties, our solution may detect the effects and consequences of certain vulnerabilities exploit or threats on the configuration of the cloud under the following conditions: (a) the vulnerability exploit or threat violates at least one of the security properties being audited, (b) the violations generate logged events and configuration data, (c) the corresponding traces of those violations in logs and configuration data are intact and not erased or tampered with, as the correctness of our audit results depends on the correct input data extracted from logs, databases, and devices.

The out-of-scope threats include attacks that do not violate the specified security properties, attacks not captured in the logs or databases, and attacks through which the attackers may remove or tamper with logged events. Existing techniques on trusted auditing may be applied to establish a chain of trust from TPM chips to auditing components, e.g., [7]).

We focus on layer 2 virtual network in this chapter, and our work is complementary to existing solutions at other network layers. We assume that not all tenants trust each other. In certain cloud offerings (e.g., private clouds), a tenant can either require not to share any physical resource with all other tenants, or provide a white (or black) list of trusted (or distrusted) tenants that he is (or not) willing to share resources with. Finally, we assume the verification results do not disclose sensitive information about other tenants and regard potential privacy issues as a future work.

Finally, we focus on auditing structural properties such as the assignment of instances to physical hosts, the proper configuration of virtualization mechanisms, and consistency of the configurations in different layers of the cloud. Those properties mainly involve static configuration information that are already stored by the cloud system at the cloud management layer and the implementation layer. The verification of operational properties, which are related to the network forwarding functionality, are out of the scope of the chapter.

4.1.3   Virtualized Cloud Infrastructure Model

In this section, we present the two-layered model that we derive to capture information related to isolated virtual networks at both the infrastructure management and the implementation layers. This model was derived based on common knowledge and studied literature on implementation and management of isolated virtual networks [20]. For instance, to elaborate and validate the infrastructure management layer model, we analyzed the abstractions exposed by the most popular cloud platforms providing tenants the capability to build virtual private networks (e.g., AWS EC2- Virtual Private Cloud (VPC) [3], Google Cloud Platform (GCP) [40], Microsoft Azure [71], VMware virtual Cloud Director (vCD) [113], and OpenStack [90]). More details will be provided in Table 4.7 (Sect. 4.5). For the implementation model, we relied on performing intensive tests on OpenStack compute and network nodes, then we supported our understanding by exploring the literature [23, 73]. Finally, we validated our two-layer model with subject matter experts.

The model allows capturing the data to be audited at each layer, its underlying semantics and relation with isolation requirements. It also defines cross-layer mappings of data in different layers to capture consistency requirements.

Infrastructure Management Model The upper model in Fig. 4.3 captures the view from the cloud infrastructure management system perspective. This layer manages virtual resources such as VMs, routers, and virtual networks (represented as entities) as well as their ownership relation (represented as relationships) with

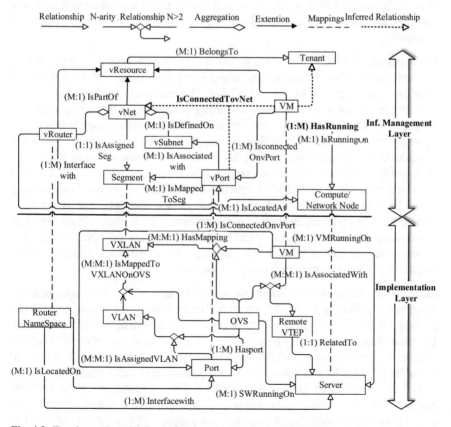

Fig. 4.3 Two-layered model for isolated multi-tenant virtualized infrastructures in the cloud: generic model for the infrastructure management layer (upper model) mapped into an implementation-specific model of the infrastructure layer (lower model)

respect to tenants. Once connected together, these resources form the tenants' virtual infrastructures. Some entities, for instance Tenant, are only maintained at the management layer and have no counterpart at the lower layer. Other entities exist across layers (e.g., VMs and ports), however, one-to-one mappings should be maintained. These mappings allow inferring missing relationships between layers and help checking consistency between the cloud-stack layers. Isolation between different virtual networks at this layer is defined using a segmentation mechanism, modeled as entity Segment. A segment should be unique for all elements of the same virtual infrastructure.

Example 4.2 Ownership is modeled using the BelongsTo relationship in Fig. 4.3 between Tenant and vResource. The related cardinality constraint (M:1) expresses that, following the directed edge, a given vResource can only belong to a single (i.e., 1) Tenant, but, a Tenant can own multiple (i.e., *M*) virtual resources. The isAssignedSeg relationship and its cardinality constraint (1:1) relating Segment to vNet allows having a unique segment per network. Relationships isConnectToVnet and HasRunningVM are of special interest to us and thus they are depicted in the model even though they can be inferred from other relationships.

Implementation Model The lower model in Fig. 4.3 captures a typical OpenStack implementation of the infrastructure management view using well-known layer 2 isolation technologies, VXLAN and VLAN. The model can capture other layer 2 isolation mechanisms such as generic routing encapsulation (GRE) by replacing the entity VXLAN with entity GRE. Some entities and relationships in this model represent the implementation of their counterparts at the management model. For instance, VXLAN combined with VLAN are implementation of entity Segment. Other entities such as virtual networking devices Open vSwitch (OVS) and Virtual Tunneling End Point (VTEP) are specific to the implementation layer as they do not exist at the infrastructure management model. They play the vital role in connecting VM instances to their hosting machines and to their virtual networks across different servers. Indeed, VTEPs are overlay-aware interfaces responsible for the encapsulation of packets with the right tunnel header depending on the destination VM and its current hosting server.

Example 4.3 At the lower model in Fig. 4.3, the ternary relationship isAssigned VLAN with cardinality (M:M:1) means that each single port in a given OVS can be assigned at most one VLAN but multiple ports can be assigned the same VLAN. To capture isolation at overlay networks spanning over different servers, the ternary relationship isMappedtoVXLAN states that each VLAN in each OVS is mapped to a unique VXLAN. The unicity between a specific port and a VLAN in an OVS as well as the unicity of the mapping of a VLAN to a VXLAN in a given OVS are inherited from the unicity of the mapping of a segment to a virtual network. The two ternary relationships hasMapping and isAssociatedWith are used to model VTEPs information existing over different physical servers. Several relations have similar semantics in both models; however, we use different names for clarity. For instance, `VMRunningOn` at the implementation layer corresponds to `isRunningOn` at the management layer.

Entities and relationships defined in these models will be used in our approach to automate the verification of isolation between tenants' virtual infrastructures. They will be essentially used to express system data and the relations among them in the form of instances of these models. Also, they will be used to express properties related to isolation as will be presented in next section.

4.2 Methodology

In this section, we detail our approach for auditing compliance of virtual layer 2 networks with respect to a multi-tenant cloud.

4.2.1 Overview

Figure 4.4 presents an overview of our approach. Our main idea is to use the derived two-layered model (Sect. 4.1) to capture the implementation of the multi-tenant virtual infrastructure along with its specification. We then verify the implementation against its specification to detect violation of the properties.

To be able to automatically process the model as the specification support for the virtual infrastructure, we first express it in first-order logic (FOL) [8]. We encode entities and relationships in both models into a set of FOL expressions, namely variables and relations. We also express isolation and consistency rules as FOL predicates based on the FOL expressions derived from the model. This process is performed offline and only once.

To obtain the implementation of the system, we collect real data from different layers (cloud management and cloud infrastructure) and use the model entities and

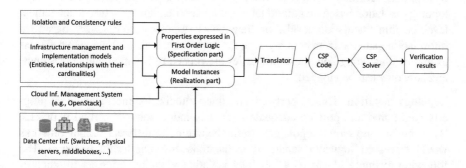

Fig. 4.4 An overview of our verification approach

relationships definitions to build an instance of the model representing the current state of the system. As we aim at detecting violations, we represent relationships between real data as instances of FOL n-ary relations without restricting instances to meet cardinality constraints. This will be detailed later on in this section. As a back-end verification mechanism, we rely on the off-the-shelf CSP solver Sugar. The latter allows formulation of many complex problems in terms of variables defined over finite domains and constraints. Its generic goal is to find a vector of values (a.k.a. assignment) that satisfies all constraints expressed over the variables. If all constraints are satisfied, the solver returns SAT, otherwise, it returns UNSAT. In the case of a SAT result, a solution to the problem, which is a specific assignment of values to the variables that satisfies the constraints, is provided. One of the key advantages of using constraint solving is to enable uniformly specifying systems data and properties in a clean formalism and covering a wide range of properties [121]. Furthermore, the latter allows to identify the data violating the verified properties as it will be explained in Sect. 4.2.3.

4.2.2   Cloud Auditing Properties

Among the goals of this work is to establish a bridge between high-level security standards and low-level implementation as well as to enable verification automation. Therefore, this section describes a set of concrete security properties related to layer 2 virtual network and overlay network isolation in a multi-tenant environment. In this chapter, we focus on the verification of structural properties gathered from the literature and the subject matter. To have a more concrete example of layer 2 virtual network isolation mechanisms, we refer to VLAN and VXLAN as examples of well-established technologies.

Table 4.1 presents an excerpt of the security properties mapped to relevant domains and control classes in security standards, namely CCM [17] (Infrastructure and virtualization security segmentation domain), ISO27017 [52] (Segregation in networks section), and NIST800 [77] (System and communications protection, System and information integrity security controls). Those properties either check topology isolation based on individual cloud layers (i.e., infrastructure management level or implementation level), or they check topology consistency based on information gathered from both layers at the same time. In the following, we discuss examples illustrating how those properties are related to isolation and consistency, and how they can be violated.

Topology Isolation This property ensures that virtualization mechanisms are properly configured and provide adequate logical isolation between virtual networks. By using isolated virtual topologies, traffic belonging to different virtual networks would travel on logically separated paths, thus ensuring traffic isolation. The following example illustrates a topology isolation violation using an instance of our model presented in Sect. 4.1.

Table 4.1 Excerpt of security properties

Category	Standard			Property		Level	
	CCM	ISO27017	NIST800	Name	Description	Mgmt.	Impl.
Topology isolation	•	•	•	Mappings unicity virtual networks-segments (P1)	Virtual networks and segments should be mapped one-to-one	×	
				Mappings unicity ports-segments (P2)	vPorts should be mapped to unique segments	×	
				Correct association ports-virtual networks (P3)	VMs should be attached to the virtual networks they are connected to through the right vPorts	×	
				Mapping unicity Ports-VLANs (P4)	Ports should be mapped to unique VLANs		×
				Mapping unicity VLANs-VXLANs (P5)	VLANs and VXLANs should be mapped one-to-one on a given server		×
				Overlay tunnels isolation (P6)	In each VTEP end, VMs are associated to their physical location and to the VXLAN assigned to the networks they are attached to		×
Topology consistency	•	•	•	VM location consistency (P7)	Consistency between VMs locations at the implementation level and at the management level	×	×
				Ports consistency (P8)	Consistency between vPorts in the implementation level and their counterparts in the management level	×	×
				Virtual links consistency (P9)	VMs should be connected to the VLANs and VXLANs in the implementation level that correspond to the virtual networks they are attached to at the management level	×	×

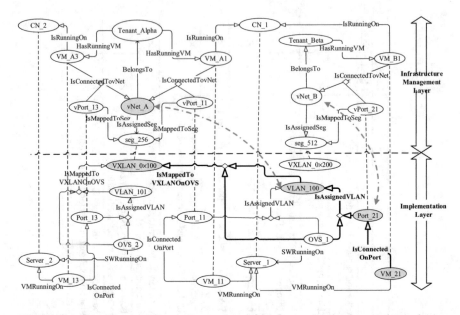

Fig. 4.5 Subsets of data and its relations at the could infrastructure implementation and management layers showing isolation violation. At the implementation level, VM_21 is connected on Port_21, that is assigned VLAN_100 as a consequence of the attack. Since VLAN_100 is mapped to VXLAN 0×100, which is mapped to seg_256 at the infrastructure management layer and the latter segment is assigned to vNet_A of Tenant_Alpha, VM_21 belonging to Tenant_Beta is now on the same network segment as VMs in vNet_A

Example 4.4 Figure 4.5 captures a subset of the data, at different layers, that is relevant to virtual networks vNet_A and vNet_B corresponding to the deployment illustrated in Figs. 4.1 and 4.2. The upper part of the figure shows a subset of the data managed by the infrastructure management layer and on the lower part, the subset of data managed by the implementation layer. Nodes represent data instances, while the directed arrows represent relations between these data instances. For example, at the infrastructure management layer, the relationship IsConnectedTovNet relates three instances of data VM_Adb, vNet_A, and vPort_11 and means that VM_Adb is connected to vNet_A on virtual port vPort_11. A cross-layer mapping, shown as small dotted undirected arrows, between some of the data instances at different layers is used to relate management-defined data to its implementation counterpart. For instance, VM_Adb and vPort_11 have each a one-to-one cross-layer mapping to VM_11 and Port_11, respectively, while no data entity at the implementation layer could be directly mapped to vNet_A at the management layer. The latter can be indirectly mapped to VXLAN_0x100 at the implementation layer via the segment seg_256. More precisely, vNet_A is implemented using VXLAN_0x100 and a set of corresponding VLANs, namely VLAN_100 and VLAN_101 (via IsMappedToVXLANonOVS), which are assigned to Port_11, Port_13, and Port_21 (via IsAssignedVLAN).

This instance of the layered model allows capturing topology isolation breaches and identifying which networks, VMs, and tenants are in this situation. Indeed VM_21 is found to be on the same virtual layer 2 segment as VM_11 and VM_13. There are two types of isolation breaches and they are illustrated as follows:

- *Intra-server topology isolation breach.* At the implementation layer, VM_21 is connected on port Port_21 (via relationship IsConnectedonPort), which is assigned VLAN_100 (via relationship IsAssignedVLAN) in the open vSwitch OVS_1. Additionally, since Port_11 connecting VM_11 is also assigned VLAN_100 on the same switch, both VM_11 and VM_21 connected via these ports are located on the same virtual network segment VLAN_100 (which corresponds to vNet_A at the infrastructure management level) leading to an isolation breach. Since both VMs are in the same server, namely Server_1, it is said to be an intra-server topology isolation at virtual layer 2. Noteworthy, without the correct mapping between VM_11 and VM_21 at the implementation layer to their respective counterparts VM_Adb and VM_Bapp1 as well as the ownership information (i.e., these VMs belong to different tenants and are connected on different virtual networks) at the management layer, we cannot conclude on the existence of this breach by only considering data from the implementation layer.

- *Inter-server topology isolation breach.* At the implementation layer, VLAN_100 that is assigned to ports Port_21 and Port_11 is mapped to VXLAN_0x100 via relationship IsMAppedToVXLANonOVS (which corresponds again to vNet_A at the infrastructure management level). However, this VXLAN identifier is also related to another VLAN_tag, namely VLAN_101, which is assigned to port Port_13 connecting VM_13 on Server_2. This is an inter-server topology isolation breach, since VM_13 and VM_21 are running on different servers (Physical Server_2 and Physical Server_1).

Topology Consistency Topology consistency consists of checking whether the topology view in the cloud infrastructure management system consistently matches the actual implemented topology, and the other way around, while considering different tenants' boundaries.

Example 4.5 (Port Consistency) Assume that a malicious insider deliberately created a port Port_40 directly on OVS_1 without passing by the cloud infrastructure management system and tagged it with VLAN_100, which is already assigned to Tenant_Alpha. This allows the malicious insider to sniff tenant's Alpha traffic on VLAN_100 via Port_40, which clearly leads to the violation of network isolation property.

4.2.3   Verification Approach

In order to systematically verify isolation and consistency properties over the model, we need to transform the model and its instances as well as the requirements into FOL expressions that can be automatically processed. In the following, we present how we express the model, the data, and the properties in FOL.

4.2.3.1   Model and Data Representation

Entities in the model are encoded into FOL variables where their domains would encompass all instances defined by the system data. Each n-ary relationship is encoded into a FOL n-ary relation over the related variables, where the instance of a given relation is the set of tuples corresponding entities-instances as defined by the relationship.

For instance, in the model instance of Fig. 4.5, the relationship IsMappedToVXLANOnOVS is translated into the following FOL relation instances capturing the actual implementation setup showing the mapping of a VLAN into a VXLAN on a given OVS instance.

- IsMapedToVXLANOnOVS(OVS_2, VLAN_101, VXLAN_0×100)
- IsMapedToVXLANOnOVS(OVS_1, VLAN_100, VXLAN_0×100)

Table 4.2 shows the main FOL relations defined in our model. These relations are required for expressing properties, which are formed as predicates as it will be presented next.

4.2.3.2   Properties Expressions

Security properties presented in Table 4.1 can be expressed as FOL predicates over FOL relations defined in Table 4.2.

Table 4.3 shows FOL predicates for the isolation properties at the infrastructure management model. Table 4.4 presents FOL predicates for the isolation properties at the implementation model. Table 4.5 summarizes the expressions of consistency-related properties.

4.2.3.3   Isolation Verification

As discussed before (Sect. 4.2.3.1), model instances are built based on the collected data and they are encoded as tuples of data representing relations' instances. On another hand, properties are encoded as predicates to specify the conditions that these relations' instances should meet.

Table 4.2 Relations in the implementation and infrastructure management models encoded in FOL

Relations	Def. at	Evaluate to $True$ if
$IsRunningOn(vm, cn)$	Mgmt.	The instance vm is located at the compute node cn
$IsMappedToSeg(vp, seg)$	Mgmt.	The virtual port vp is mapped to the segment seg
$IsAssignedSeg(vNet, seg)$	Mgmt.	The virtual network $vNet$ is assigned the segment seg
$IsConnectedTovNet(vm, vNet, vp)$	Mgmt.	vm is connect to $vNet$ on the virtual port vp
$HasPort(sw, p)$	Impl.	The virtual switch sw has a port p
$IsAssignedVLAN(sw, p, vlan)$	Impl.	The port p on switch sw is assigned the VLAN $vlan$
$IsMappedToVXLANOnOVS(sw, vlan, vxlan)$	Impl.	$vlan$ is mapped to $vxlan$ on the virtual switch sw
$SwRunningOn(sw, s)$	Impl.	The switch sw is running on the server s
$VMRunningOn(vm, s)$	Impl.	The VM vm is running on the server s
$IsConnectedOnPort(vm, sw, p)$	Impl.	The VM vm is connected on port p belonging to the switch sw
$HasMapping(ovs, vm, vxlan)$	Impl.	The VM vm is associated to $vxlan$ on a remote switch ovs
$IsAssociatedWith(ovs, vm, vtep)$	Impl.	The VM vm is associated to the remote VTEP $vtep$ on ovs
$IsRelatedTo(vtep, s)$	Impl.	the VTEP $vtep$ is defined on the server s

To verify the security properties, we use both properties' predicates and relations' instances to formulate the CSP constraints to be fed into the CSP solver. Since CSP solvers provide solutions only in case the constraint is satisfied (SAT), we define constraints using the negative form of the FOL predicates presented in Tables 4.3, 4.4, and 4.5. Hence, the solution provided by the CSP solver gives the relations' instances for which the negative form of the property is satisfied, meaning that a violation has occurred.

To better explain how the CSP solver allows to obtain the violation evidence, we provide hereafter an example of the verification of the inter-server isolation property provided in Example 4.4.

Example 4.6 We assume that VM location consistency and port consistency properties were verified to be met by the configuration. From the infrastructure man-

Table 4.3 Isolation properties at the infrastructure management level in FOL

Properties	FOL expressions
Mappings unicity virtual networks and segments (P1)	$\forall vNet1, vNet2 \in vNET, \forall seg1, seg2 \in Segment$: $[IsAssignedSeg(vNet1, seg1) \wedge$ $IsAssignedSeg(vNet2, seg2) \wedge$ $\neg(vNet1 = vNet2) \rightarrow \neg(seg1 = seg2)] \wedge$ $[IsAssignedSeg(vNet1, seg1) \wedge$ $IsAssignedSeg(vNet2, seg2) \wedge$ $\neg(seg1 = seg2) \rightarrow \neg(vNet1 = vNet2)]$
Mappings unicity Ports-Segments (P2)	$\forall seg1, seg2 \in Segment, \forall vp \in vPORT$: $IsMappedToSeg(vp, seg1) \wedge$ $IsMappedToSeg(vp, seg2) \rightarrow (seg1 = seg2)$
Correct association ports-virtual networks (P3)	$\forall vm \in VM, \forall vNet \in vNET, \forall seg1, seg2 \in$ $Segment, \forall vp \in vPort$: $IsConnectedTovNet(vm, vNet, vp) \wedge$ $IsAssignedSeg(vNet, seg1)$ $\wedge IsMappedToSeg(vp, seg2) \rightarrow (seg1 = seg2)$

Table 4.4 Isolation properties at the implementation level in FOL

Properties	FOL Expressions
Mapping unicity Ports-VLANs (P4)	$\forall sw \in OVS, \forall p \in Port, \forall vlan1, vlan2 \in VLAN$: $HasPort(sw, p) \wedge$ $IsAssignedVLAN(sw, p, vlan1) \wedge$ $IsAssignedVLAN(sw, p, vlan2) \rightarrow$ $(vlan1 = vlan2)$
Mapping unicity VLANs-VXLANs (P5)	$\forall vxlan1, vxlan2 \in VXLAN, \forall vlan \in vlan, \forall sw \in$ $OVS,$ $\forall p \in PORT$: $(IsAssignedVLAN(sw, p, vlan) \wedge$ $IsMappedToVXLANOnOVS(sw, vlan, vxlan1) \wedge$ $IsMappedToVXLANOnOVS(sw, vlan, vxlan2) \rightarrow$ $(vxlan1 = vxlan2)$
Overlay tunnels isolation (P6)	$\forall vm \in VM, \forall sw1, sw2 \in OVS, \forall p \in$ $PORT, \forall vxlan1, vxlan2 \in VXLAN,$ $\forall s1, s2 \in Server, \forall vtep \in RemoteVTEP, \forall vlan \in$ $VLAN$: $HasPort(sw1, p) \wedge SWRunningOn(sw1, s1) \wedge$ $IsConnectedOnPort(VM, sw1, p) \wedge$ $IsAssignedVLAN(sw1, p, vlan) \wedge$ $IsMappedToVXLANOnOVS(sw1, vlan, vxlan1) \wedge$ $IsAssociatedWith(sw2, vm, vtep) \wedge$ $HasMapping(sw2, vm, vxlan2) \wedge$ $IsRelatedTo(vtep, s2) \rightarrow (s1 = s2) \wedge (vxlan1 = vxlan2)$

Table 4.5 Topology consistency properties in FOL

Properties	FOL expressions
VM location consistency (P7)	$\forall vm1 \in VM, \forall cn \in COMPUTEN:$ $IsRunningOn((vm1, cn)) \rightarrow$ $\exists vm2 \in iVM, \exists s \in SERVER:$ $VMRunningOn(vm2, s) \wedge$ $(vm1 = vm2) \wedge (cn = s)$
Ports consistency (P8)	$\forall vNet \in vNET, \forall seg \in Segment, \forall vp \in vPORT:$ $IsAssignedSeg(vNet, seg) \wedge$ $IsMappedToSeg(vp, seg) \rightarrow$ $[\exists sw \in OVS, \exists vxlan \in VXLAN, \exists vlan \in VLAN,$ $\exists p \in PORT : IsAssignedVLAN(sw, p, vlan)$ $\wedge IsMappedToVXLANOnOVS(sw, vlan, vxlan) \wedge$ $(seg = vxlan)(vp = p)]$
Virtual links consistency (P9)	$\forall vm1 \in iVM, \forall vxlan \in VXLAN, \forall sw \in$ $OVS, \forall vlan \in VLAN, \forall p \in PORT:$ $IsConnectedOnPort(vm1, sw, p) \wedge$ $IsAssignedVLAN(sw, p, vlan) \wedge$ $IsMappedToVXLANOnOVS(sw, vlan, vxlan) \rightarrow$ $[\exists vm2 \in vVM, \exists vNet \in vNET, \exists seg \in$ $Segment, \exists vp \in vPORT:$ $IsConnectedTovNet(vm2, vNet, vp) \wedge (vm1 =$ $vm2) \wedge$ $IsAssignedSeg(vNet, seg) \wedge (seg = vxlan)]$

agement level, we recover the virtual networks connecting each VM and their corresponding segment. This is captured through the following relation instances:

- `IsConnectedTovNet((VM_Bapp1, vNet_B, vPort_21),` `(VM_Adb, vNet_A, vPort_11), (VM_Aweb, vNet_A,` `vPort_13))`
- `IsAssignedSeg((vNet_B, seg_512),(vNet_A, seg_256)))`

From the implementation level, we recover the OVS and the ports connecting VMs in addition to their assigned VLAN tags and VXLAN identifiers captured through the following relation instances:

- `IsConnectedOnPort((VM_21, OVS_1, Port_21),(VM_11,` `OVS_1, Port_11), (VM_13, OVS_2, Port_13))`
- `IsAssignedVLAN((OVS_1, Port_21, vlan_100),(OVS_1,` `Port_11, vlan_100), (OVS_2, Port_13, vlan_101))`
- `IsMappedToVXLANOnOVS((OVS_1, vlan_100, vxlan_0×100),` `(OVS_2, vlan_101, vxlan_0×100))`

We would like to verify that the VXLAN identifier assigned to a virtual network at the implementation level is equal to the segment assigned to this same network at the infrastructure management level (after conversion to decimal), which is expressed by *virtual link consistency* property (*P9*). To find whether there exist relations' tuples that falsify this property ($\neg P9$), we first formulate the CSP instance using the negative form of the corresponding predicate, which corresponds to the following predicate:

$$\neg P11 = \exists vm1 \in iVM, \exists vxlan \in VXLAN, \quad (4.1)$$

$$\exists sw \in OVS, \exists vlan \in VLAN, \exists p \in PORT,$$

$$\forall vm2 \in vVM, \forall vNet \in vNET, \forall seg \in Segment, \forall vp \in vPORT:$$

`IsConnectedOnPort(vm1, sw, p)` $\wedge$ `IsAssignedVLAN(sw, p, vlan)`$\wedge$

`IsMappedToVXLANOnOVS(sw, vlan, vxlan)`$\wedge$

$\neg$`IsConnectedTovNet(vm2, vNet, vp)` $\vee$ $\neg$`(vm1 = vm2)`$\vee$

$\neg$`IsAssignedSeg(vNet, seg)` $\vee$ $\neg$`(seg = vxlan)`

By verifying predicate (4.1) over all the aforementioned relations' instances, the solver finds an assignment such that the above predicate becomes true, which means that the property *P9* is violated. The predicate instance that caused the violation can be written as follows:

$$\text{IsConnectedOnPort(VM\_21, OVS\_1, Port\_21)} \wedge \quad (4.2)$$

`IsAssignedVLAN(OVS_1, Port_21, vlan_100)`$\wedge$

`IsMappedToVXLANOnOVS(OVS_1, vlan_100, vxlan_0 × 100)`$\wedge$

$\neg$`IsConnectedTovNet(VM_Bapp1, vNet_B, vPort_21)` $\vee$ $\neg$`(VM_21 = VM_Bapp1)`$\vee$

$\neg$`IsAssignedSeg(vNet_B, seg_512)` $\vee$ $\neg$`(seg_512 = vxlan_0 × 100)`

Since `seg` is equal to `512` and the decimal value of VXLAN0_×100, namely `vxlan`, is `256`, then the equality `seg=vxlan` will be evaluated to false and $\neg$`(seg=vxlan)` will be evaluated to true, which makes the assignment in predicate (4.2) satisfying the constraint. This set of tuples provides the evidence about what values breached the security property *P9*. Note that as VM consistency and port consistency properties were assumed to be verified, the equality between VM_Bapp1 and VM_21 holds (based on their identifiers that could be their MAC addresses for instance).

In the following section, we present our auditing solution integrated into OpenStack and show details on how we use the CSP solver Sugar as a back-end verification engine.

4.3 Implementation

In this section, we briefly review the OVS-based networking, we then detail our implementation and its integration into OpenStack and Congress [89], an open-source framework implementing policy as a service for OpenStack.

4.3.1 Background

As we are interested in auditing the infrastructure virtualization and network segregation, we first investigated OpenStack documentation to learn which services are involved in the creation and maintenance of the virtual infrastructure and networking. We found that Nova and Neutron services in OpenStack are responsible in managing virtual infrastructure and networking at the management layer. We also investigated the implementation level and found that OVS instances running in different compute nodes are the main components that implement the virtual infrastructure.

In our settings, an OVS defines two interconnected bridges, the integration bridge (br-int) and the tunneling bridge (br-tun). VMs are connected via a virtual interface (tap device)[3] to br-int. The latter acts as a normal layer 2 learning switch. It connects VMs attached to a given network to ports tagged with the corresponding VLAN, which ensures traffic segregation inside the same compute node.

Each tenant's network is assigned a unique VXLAN identifier over the whole infrastructure. The *br-tun* is endowed with OpenFlow rules [80] that map each internal VLAN-tag to the corresponding VXLAN identifier and vice versa. For egress traffic, the OpenFlow rules strip the VLAN-tag and set the corresponding VXLAN identifier in order to transmit packets over the physical network. Conversely, for ingress traffic, OpenFlow rules strip the VXLAN identifier from the received traffic and set the corresponding VLAN-tag.

4.3.2 Integration into OpenStack

We mainly focus on four components in our implementation: the data collection engine, the data processing engine, the compliance verification engine, and the dashboard and reporting engine. In the following, we describe our implementation details.

---

[3]This direct connection is an abstraction of a chain of one-to-one connections from the virtual interface to the *br-int*. In fact, the tap device is connected to the Linux bridge *qbr*, which is in turn connected to the br-int.

Data Collection Engine The data collection engine involves several components
of OpenStack, e.g., Nova and Neutron for collecting audit data from databases
and log files, different policy files, and configuration files from the OpenStack
ecosystem, and configurations from various virtual networking components such as
OVS instances in all physical servers to fully capture the configuration and virtual
networks state. We present hereafter different sources of data along with the current
support for auditing offered by OpenStack and the virtual networking components.
Sample data sources are shown in Table 4.6. We use different sources including
OpenFlow tables extracted from OVS instances in every compute node, and Nova
and Neutron databases:

- *OpenStack.* We rely on a collection of OpenStack databases, that can be read
  using component-specific APIs. For instance, in Nova database, table *Instance*
  contains information about the project (tenant) and the hosting machine, table
  Migration contains migration events' related information such as the source
  compute and the destination compute. The Neutron database includes various
  information such as port mappings for different virtualization mechanisms.
- *OVS.* OpenFlow tables and internal OVS databases in different compute nodes
  constitute another important source of audit data for checking whether there
  exists any discrepancy between the actual distributed configuration at the implementation layer and the OpenStack view.

Table 4.6 Sample data sources in OpenStack, open vSwitch and tenants' requirements

Relations	Sources of data
IsRunningOn	Table *Instances* in Nova database
IsAssignedSeg	Table *ml2_network_segments* in Neutron database
IsMappedToSeg	Table *neworkconnections* in Neutron database
IsConnectedTovNet	Table *Instances* in Nova database
HasPort	OVS instances located at various compute nodes, *br_int* configuration
IsAssignedVLAN	OVS instances located at various compute nodes, *br_int* configuration
IsMappedToVXLANOnOVS	OVS instances located at various compute nodes, *br_tun* OpenFlow tables
VMRunningOn	OVS instances located at various compute nodes, *br_int* configuration
SWRunningOn	The infrastructure deployment
IsConnectedOnPort	OVS instances located at various compute nodes, *br_int* configuration
HasMapping	OVS instances located at various compute nodes
IsAssociatedWith	OVS instances located at various compute nodes
IsRelatedTo	OVS instances located at various compute nodes
DoesnotTrust	The tenant physical isolation requirement input

For the sake of comprehensiveness in the data collection process, we firstly check fields of a variety of log files available in OpenStack, different configuration files, and all Nova and Neutron database tables. We also debug configurations of all OVS instances distributed over the compute nodes using various OVS's utilities. Mainly, we recovered ports' configurations (e.g., ports and their corresponding VLAN tags) from the integration bridges using the utility `ovs-vsctl show`, and we extracted VLAN-VXLAN mappings form the tunneling bridges' OpenFlow tables using `ovs-ofctl dump-flows`. The tunneling bridge maintains a chain of OpenFlow tables for handling ingress and egress traffic. In order to recover the appropriate data, we identify the pertinent tables where to collect the VLAN-VXLAN mappings from. Through this process, we identify all possible types of data, their sources, and their relevance to the audited properties.

Data Processing Engine The data processing engine, which is implemented in Python and Bash scripts, mainly retrieves necessary information from the collected data according to the targeted properties, recovers correlation from various sources, eliminates redundancies, converts it into appropriate formats, and finally generates the source code for Sugar.

- Firstly, based on the properties, our plug-in identifies the involved relations. The relations' instances are either fetched directly from the collected data such as the support of the relation `BelongsTo`, or recovered after correlation, as in the case of the relation `IsConnectedTovNet`.
- Secondly, our processing plug-in formats each group of data as an n-tuple, i.e., `(resource, tenant)`, `(ovs, port, vlan)`, etc.
- Finally, our plug-in uses the n-tuples to generate the portions of Sugar's source code, and append the code with the variable declarations, relationships, and predicates for each security property.

Checking consistent topology isolation in virtualized environments requires considering configurations generated by virtualization technologies at various levels, and checking that mappings are properly maintained over different layers. OpenStack maintains tenants' provisioned resources but does not maintain overlay details of the actual implementation. Conversely, current virtualization technologies do not allow mapping VMs, networks, and traffic details to their owners. Therefore, we map virtual topology details at the implementation level to the corresponding tenant's network to check whether isolation is achieved at this level. Here are examples of mappings to provide per-tenant evidences for resources and layer 2 virtual network isolation. Figure 4.6 relates relations of property *P9* along with some of their data support to their respective data sources.

- At the OpenStack level, tenants' VMs are connected to networks through subnets and virtual ports. Therefore, we correlate data collected from *Insatances* Nova table to recover a direct connection between VMs and their connecting networks at the centralized view through the relation `IsConnectTovNet`. We also keep track of their owners.

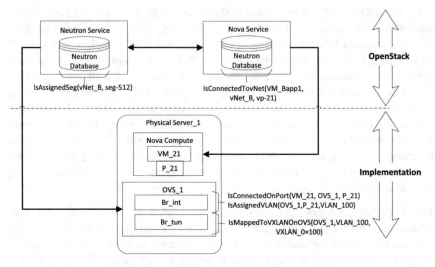

Fig. 4.6 Mapping of relations involved in property $P9$ to their data sources

- At the virtualization layer, networks are identified only through their VXLAN identifiers. We map each network's segment identifier recovered from OpenStack (Neutron Database) to the VXLAN identifier collected from OVS instances (br_tun OpenFlow tables) to be able to map each established flow to the corresponding networks and tenants. Furthermore, for each physical server, we assign VMs to the ports that they are connected to through the relation IsConnectOnPort, and we assign ports to their respective VLAN-tags through the relation IsAssignedVLAN from the configurations details recovered from br-int configuration in OVS.
- At the OpenStack level, ports are directly mapped to segment identifiers, whereas at the OVS level, ports are mapped to VLAN-tags and mappings between the VLAN-tags and VXLAN identifiers are maintained in OpenFlow tables distributed over multiple OVS instances. To overcome this limit, we devised a script that recovers mappings between VLAN-tags and the VXLAN identifiers from the flow tables in br-tun using the ovs-ofctl command line tool. Then, it recovers mappings between ports and VLAN-tags from the OpenvSwitch database using the ovs-vsctl command line utility.

Depending on the properties to be checked, our data processing engine encodes the involved instances of the virtualized infrastructure model as CSP variables with their domains definitions, where instances are values within the corresponding domain. The CSP code mainly consists of four parts:

- *Variable and domain declaration.* We define different entities and their respective domains. For example, *TENANT* is defined as a finite domain ranging over integer such that *(domain* TENANT 0 max_tenant is a declaration of a domain of tenants, where the values are between 0 and max_tenant.

- *Relation declaration.* We define relations over variables and provide their supports (instances) from the audit data. Relations between entities and their instances are encoded as relation constraints and their supports, respectively. For example, `HasRunningVM` is encoded as a relation, with a support as follows: `(relation HasRunningVM 2 (supports (vm1,t1) (vm2,t2)))`, where the support of this relation (e.g., $(vm1, t1)$) will be fetched and pre-processed in the data processing step.
- *Constraint declaration.* We define the negation of each property in terms of predicates over the involved relations to obtain a counterexample in case of a violation.
- *Body.* We combine different predicates based on the properties to verify using Boolean operators.

Compliance Verification The compliance verification engine performs the verification of the properties by feeding the generated code to Sugar. Finally, Sugar provides the results on whether the properties hold or not. It also provides evidence in case of non-compliance.

Example 4.7 In this example, we discuss how our auditing framework can detect the violation of the virtual links inconsistency caused by the inter-compute node isolation breach described in Example 4.4.

Firstly, our program collects data from different sources. Then, the processing engine correlates and converts the collected data and represents it as tuples; for an example: `(18045 6100 21)` `(6100 512)` reflect the current configuration at the infrastructure management level, and `(18045 1 21)` `(1 21 100)` `(1 100 256)` correspond to a given network's configuration at the implementation level, where `VM_Bapp1: 18045`, `VM_21: 18045`, `vNet_B: 6100`, `seg_512: 512`, `vPort_21: 21`, `OVS_1: 1`, `Port_21: 21`, `VLAN_100: 100`, `vxlan_1×100: 256`. Additionally, the processing engine interprets each property and generates the associated Sugar source code (see Listing 4.1 for an excerpt of the code) using processed data and translated properties. Finally, Sugar is used to verify the security properties.

The predicate *P9* for verifying *virtual link consistency* evaluates to true if there exists a discrepancy between the network `VM_Bapp1` is connected to according to the infrastructure management view, and the layer 2 virtual network `VM_Bapp1` is effectively connected to at the implementation level. In our case, the predicate evaluates to true since `vxlan0×100`≠`seg_512` (as detailed in Example 4.6), meaning that `VM_Bapp1` is connected on the wrong layer 2 virtual network.

Listing 4.1 Sugar Source Code

```
// Declaration
(domain iVM 0 100000)(domain OVS 0 400)(domain PORT 0 100000)
(domain VLAN 0 10000) (domain VXLAN 0 10000)(doamin vVM 0 100000)
(domain VNET 0 10000) (domain SEGMENT 0 10000)(domain VPORT 0 100000)
( int  vm1 iVM) (int  vm2 vVM)(int sw OVS) (int p PORT)(int vlan  VLAN)
( int  vxlan  VXLAN)(int vnet VNET) (int seg SEGMENT) (int vp VPORT)
```

```
// Relations Declarations and Audit data as their support from the
   infrastructure  management level
( relation  IsConnectedTovNet 3   (supports  (18045 6100 21)  (18037 6150 7895)
(18038 6120 2566)  (  18039 6230 554)(18040 6230 4771)(966)))
( relation  IsAssignedSeg 2  (supports  (6150 356)(6120 485)(6230 265)
(6100 512)(6285 584)(6284 257)))
// Relations  Declarations  and Audit data as  their  support from the
   implementation  level
( relation  IsConnectedOnPort 3  (supports(((18045  1  21)(18037 96 23)
(18046 65 32)(18040 68 8569)(18047 78954))
( relation  IsAssignedVLAN 3 (supports(92 13 41)(92 14 42)(85 38 11)))
( relation  IsMAppedToVXLANOnOVS 3 (supports (1 100 256)(92 6018 9)
(92  6019 10)))
// Security  properties  expressed  in terms of  predicates  over  relation
   constraints
( predicate  (P  vm1 vm2 vnet seg vxlan   sw p vp)
(and(IsConnectedOnPort vm1 sw p)(IsAssignedVLAN sw p vlan)
(IsMappedToVXLANOnOVS sw vlan vxlan)(IsConnectedTovNet vm2 vnet vp)
(IsAssignedSeg vnet  seg)(eq vm1 vm2)(not(eq seg  vxlan))))
// The body
(P  vm1 vm2 vnet seg vxlan   sw p  vp)
```

Understanding Violations Through Evidences As explained in Sect. 4.2.3.3, we define constraints using the negative form of properties' predicates. Thus, if a solution satisfying the constraint is provided by the CSP solver, then the latter solution is a set of variable values that make the negation of the predicates evaluate to true. Those values indicate the relation instances (system data) that are at the origin of the violation; however, they might be unintelligible to the end users. Therefore, we replace the variables' numerical values by their high-level identifiers, which would help admins identify the root cause of the violation and fix it eventually.

Example 4.8 From Example 4.7, the CSP solver concludes that the negative form of the property is satisfied, which indicates the existence of a violation. Furthermore, the CSP solver outputs the following variable values as an evidence: $vm1=18045$, $vm2=18045$, $vnet=6100$, $seg=512$, $vxlan=100$, $sw=1$, $p=21$, $vp=21$. To make the evidence easier to interpret, we replace the value 6100 of the variable vnet by $vNet\_B$, the value 18045 of the variable vm2 by VM_Bapp1, and the value 21 of the variable vp by vp_21. Using this information, the admin will conclude that VM_Bapp1 is connected to another $Tenant\_Alpha$'s layer 2 virtual network at the implementation level identified through $VXLAN\_0 \times 100$.

Dashboard and Reporting Engine We further implement the web interface (i.e., dashboard) in PHP to place verification requests and display verification reports. In the dashboard, tenant admins are initially allowed to select different standards (e.g., ISO 27017, CCM V3.0.1, NIST 800-53, etc.). Afterwards, security properties under the selected standards can be chosen. Once the verification request is placed, the summarized verification results are shown in the verification report page. The details of any violation with a list of evidences are also provided.

4.3.3 Integration into OpenStack Congress

To demonstrate the service agnostic nature of our framework, we further integrate our system with the OpenStack Congress service [89]. Congress implements policy as a service in OpenStack in order to provide governance and compliance for dynamic infrastructures. Congress can integrate third party verification tools using a data source driver mechanism [89]. Using Congress policy language that is based on Datalog, we define several tenant-specific security policies. Then, we use our processed data to detect those security properties for multiple tenants. The outputs of the data processing engine are provided as input for Congress to be asserted by the policy engine. This allows integrating compliance status for some policies whose verification is not yet supported by Congress.

4.4 Experiments

In this section, we evaluate scalability of our approach by measuring the response time of the verification task as well as the CPU and memory consumption for different sizes of cloud and in different scenarios (a breach violates some properties or no breach).

4.4.1 Experimental Setting

We set up a real environment including 5 tenants, 10 virtual networks each having 2 subnets, 10 routers, and 100 VMs. We utilize OpenStack Mitaka with one controller and three compute nodes running Ubuntu 14.04 LTS. The controller is empowered with two Intel Xeon E3-1271 CPU and 4 GB of memory. Each compute node benefits from one CPU and 2 GB of memory. To further stress the verification engine and assess the scalability of our solution, we generated a simulated environment including up to 6k virtual networks and 60K VMs with the ratio of 10 VMs per virtual network. As a back-end verification tool, we use the CSP solver Sugar V2.2.1 [108]. All the verification experiments are run on an Amazon EC2 C4.Large Ubuntu 16.04 machine (2 vCPU and 3.75 GB of memory).

4.4.2 Results

Experimental results for *physical isolation*, *virtual resources isolation*, and *port consistency* properties are reported in Chap. 3. In addition, we consider three additional properties from Table 4.1, where each is selected from one of the three categories defined therein. Thus, we consider for the experiments the following three properties, one from each category:

- *Mapping unicity virtual networks-segments* (P1), which is a topology isolation property checked at the infrastructure management level.
- *Mapping unicity VLANs-VXLANs* (P5), which is a topology isolation property checked at the implementation level.
- *Virtual links consistency* (P9), which checks that a VM is connected to the right VXLAN at the implementation level.

In the first set of experiments, we design two configuration scenarios to study different response times in two possible cases: presence of violations and absence of violations. This is because the verification of these two scenarios is expected to have different response times due to the time required to find the evidence of the violation.

In the first scenario, we implement in our environment a configuration of the virtual infrastructure where none of the studied properties are violated. In the second scenario, we implement the topology isolation attack described in Example 4.4. For the latter scenario, as generally, a fast yes or no answer on the compliance status of the system is required by the auditor, we only consider the response time to report evidence for the first breach. Note that we do not report the average response time to find all compliance breaches as this depends on the number of breaches, their percentage to the total input size, and their distribution in the audit information. Meanwhile, as the real-life scenarios can dramatically vary from one environment to another, we cannot use any average number, percentage, or distribution of compliance breaches applying to all possible use cases. Therefore, we present in Fig. 4.7, the verification time for no security breach detected (left side chart) and the verification time to report non-compliance and provide evidence for the first security breach (right side chart) for different datasets varying from 5K up to 60K VMs. Note that, we implement the attack scenario of topology isolation described in Example 4.4 by randomly modifying some VLAN ports and VLAN to VXLAN mappings.

As indicated in the left chart of Fig. 4.7, the time required for verifying P1 and P5, where there is no breach, is 0.6 and 4.5 s, respectively, for the largest dataset of 60K VMs. The verification time for those properties increases linearly and smoothly when the size of the cloud infrastructure increases and there is no breach. However, the verification time for property P9 is 102 s for 30k VMs and 581 s for 60k VMs. The difference in response time for P9 is justified as the latter is more complex than other properties and involves more relations and thus larger input data. Later in this section, we will show how one can decrease the response time for the verification of P9 to get more acceptable boundaries.

According to Fig. 4.7 (right side chart), the time required to find the first breach and build the supporting evidence for each one of the three properties remains under 5 s for the largest dataset, which is two orders of magnitude smaller than the time required to assert compliance for the entire system. The time required to find the first breach depends on several factors such as the predicates affected by the breach

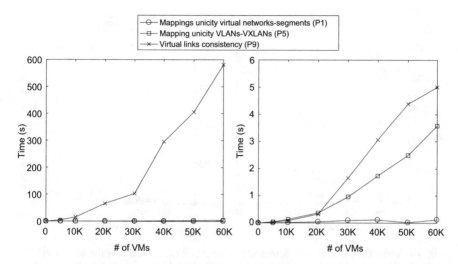

Fig. 4.7 Verification time as a function of number of VMs for properties P1, P5, and P9: (left side) time to report no breach of compliance, and (right side) time to find the first breach and build evidence of non-compliance

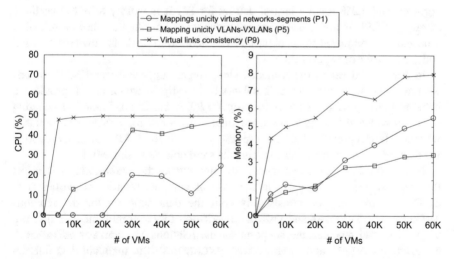

Fig. 4.8 CPU (left side) and memory (right side) usage to verify no-compliance breach for properties P1, P5 and P9

and the location of the breach in the input file. However, the latter response time is always shorter than the time required for asserting the compliance of the system.

The left side chart of Fig. 4.8 reports CPU consumption percentage as a function of the datasets' size, up to 60k VMs. For the largest dataset, the peak CPU usage reaches 50% for P9 and does not exceed 25% for P1. Also, the highest memory usage observed does not exceed 8% for P9 verification (see the right chart of

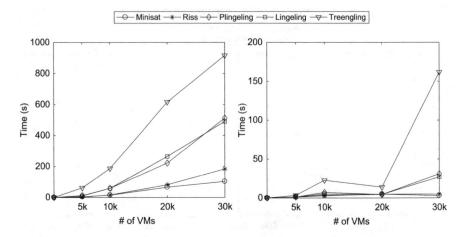

Fig. 4.9 Verification time using different SAT solvers for P9 as a function of the number of VMs: (left side) time to report no breach of compliance, and (right side) time to find the first breach and build evidence of non-compliance

Fig. 4.8), and 3.3% for the largest dataset for P5. It is worthy to note that these amounts of CPU/memory usage are not monopolized during the whole verification time and they represent the peek usage. We therefore remark the low cost on CPU and memory for our approach.

In our second set of experiments, since Sugar supports several SAT solvers, we run Sugar with different SAT solvers to investigate which option provides a better response time, particularly for property P9. According to Fig. 4.9, Treengling solver provides the longest response time with 900 s for a 30k VMs dataset, whereas Minisat provides the best response time with 102 s. All previously reported verification results in the other experiments were obtained using Minisat.

In our third set of experiments, we investigate the parameters that affect the response time, particularly in the case of complex security properties such as P9. To this end, we consistently split the data supports for the relations IsConnectedToVnet and IsAssignedSeg of P9 over multiple CSP files (up to 16 files) and repeated the supports for the relations IsConnectedOnPort, IsAssignedVLAN, and IsMappedToVXLANOnOVS to maintain data interdependency. Figure 4.10 reports the response times for the parallel verification of different CSP sub-instances of P9 using multiple processing nodes for the largest dataset (60K VMs). By splitting the data support into two CSP files, the verification time already decreases from 581 to 168 s (i.e., a factor of improvement of 71%), whereas it decreases up to 4.6 s when splitting the data over 16 CSP sub-instance files.

Based on this last experiment, we can conclude that splitting the input data for the same property to be verified using parallel instances of CSP solvers can improve the response time. However, this should be performed while considering the dependency between different relations and their supports in the predicate to be solved.

Fig. 4.10 Verification time as function of the number of processing nodes for P9 for a dataset of 60k VMs, where each processing node verifies a separate CSP sub-instance of P9

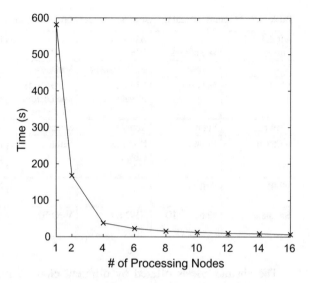

Based on those results, we conclude that our solution provides acceptable response time for auditing security isolation in the cloud, particularly, in the case of off-line auditing. While the verification of simple properties is scalable for large cloud virtual infrastructures, response time for complex properties involving large input data can induce more delays that can be still acceptable for auditing after the fact. However, response time for those properties can be considerably improved by splitting their CSP instance into sub-instances involving smaller amounts of data to be checked in parallel. Note that our analysis holds for the specific scenario where security properties are expressed as constraints defined as logical operations over relations, which is only a subset of possible constraints that can be offered by the CSP solver Sugar (the complete set of constraints supported by Sugar can be found in [109]). Expressing new security properties with other kinds of constraints may require performance to be reassessed through new experiments.

4.5   Discussion

The experimental results presented in the previous section show that CSP solvers can be used for off-line auditing verification with acceptable response time and scalability in case of moderate size of data. Our results also show that for properties handling larger datasets, we need to decompose the verification of the properties over smaller chunks of data to improve the response time. Additionally, we explore a parallel processing approach to improve the response time for very large datasets. Note that the response time can be further improved to achieve on-line auditing by improving the performance of the CSP-solving phase [70], which is an interesting future direction.

Table 4.7 Mapping virtual infrastructure model entities into different cloud platforms

Model entities	OpenStack	AWS-EC2-VPC	GCP	Microsoft Azure	VMware vCD
VM	Instance	EC2 instance	VM instance	Azure VM	VM
vNet	Network	Virtual private cloud	Auto mode vpc custom mode vpc	Virtual network	Network
vSubnet	Subnet	Subnet	Subnet	Subnet	Subnet
vRouter	Router	Routing tables	Routes	BGP and user-defined routes	Distributed logical routers
vPort	Port	–	–	NIC	Port/port-group
Segment	Network ID	VPC ID	VPC ID	Virtual network ID	Network ID

The abstract views offered by different cloud platforms to tenants are quite similar to what we propose at the cloud infrastructure management view of our model. For instance, both Amazon AWS EC2-VPC (Virtual Private Cloud) [3], Google Cloud Platform (GCP) [40], Microsoft Azure [71], and VMware virtual Cloud Director (vCD) [113] provide tenants with the capability to create virtual network components as software abstractions, enabling to provision virtual networks. Therefore, our model can capture the main virtual components that are common to most of the IaaS management systems with minor changes. Table 4.7 maps the entities of our infrastructure management view model to their counterparts in the cloud platforms cited above.

Eucalyptus [32] is an open source IaaS management system. The Eucalyptus virtual private cloud (VPC) is implemented with MidoNet [72], an open-source network virtualization platform. In the same fashion as OpenStack Neutron, Eucalyptus MidoNet supports virtualization mechanisms such as VLAN and VXLAN to implement large scale layer 2 virtual networks spanning over the cloud infrastructure. Therefore, our implementation layer model can be applied to Eucalyptus implementations with minor changes.

However, implementation details may significantly vary between different platforms. Furthermore, cloud providers typically do not disclose their implementation details to their customers. Therefore, the implementation layer of our model along with the extracted properties might need to be revised according to the implementation details of each cloud deployment if those are provided. However, this needs to be done only once before initializing the compliance auditing process.

Our current solution is designed for the specific OpenStack virtual layer 2 implementation mainly relying on VLAN and VXLAN as well-established network virtualization technologies, and OVS as a widely used virtual switch implementation. However, as we use high-level abstractions to represent virtual layer 2 connectivity and tunneling technologies, we believe that our approach remains applicable in case of other overlay technologies such as GRE. In small to medium clouds, where VLAN tags are sufficient to implement all layer 2 virtual networks on top of the physical network, our implementation model is simplified and the security properties related to the mapping between VLAN and VXLAN can be skipped.

Among the main advantages of using a CSP solver for the verification is that it allows to integrate new audit properties with a minor effort. In our case, including a new property consists of expressing it in FOL and identifying the audit data this property should be checked against. These properties can be modified at any stage of the cloud life cycle and including them into the verification process can be decided depending on the cloud deployment offering (e.g., public or private cloud).

In this work, we extracted a set of security properties from specific domains in relevant cloud security standards that are mainly related to infrastructure virtualization and tenants' networks isolation (e.g., Infrastructure Virtualization Systems domain from CCM, and Segregation in Networks section from ISO27017). Thus, our list of implemented security properties is not meant to exhaustively cover the entire security standards. Covering other security control classes for the standards requires extracting new sets of security properties to be modeled and formalized. However, as we handle general concepts for modeling different virtual resources, we believe that our approach can be generalized to other security properties to support the entire security standards.

Finally, through this work, we show the applicability and the benefit of our formal approach in verifying security properties while providing evidences to assist admins finding the root causes of violations. As discussed in this section, we believe our high-level abstractions-based model can be easily mapped to different cloud platforms. However, the model needs to be adapted to support those different cloud platforms' implementation details, and augmented to support new security properties.

4.6 Conclusion

Auditing compliance of the cloud with respect to security standards faces several challenges. In this chapter, we presented an automated off-line auditing approach while focusing on verifying network isolation between tenants' virtual networks in OpenStack-managed cloud at layer 2 and overlay. As it was shown in this chapter, the layered nature of the cloud stack and the dependencies between layers make existing approaches that separately verify each single layer ineffective. To this end, we devised a model that captures for each cloud-stack layer, namely the infrastructure management and the implementation layers, the virtual network entities along with their inter-dependencies and their isolation mechanisms. The model helped in identifying the relevant data for auditing network isolation and capturing its underlying semantics across multiple layers. Furthermore, we devised a set of concrete security properties related to consistent network isolation on virtual layer 2 and overlay networks to fill the gap between the standards and the low-level data. To provide a reliable and evidence-based auditing, we encoded properties and data as a set of constraints satisfaction problems and used an off-the-shelf CSP solver to identify compliance breaches. Our approach furthermore pinpoints the roots of breaches enabling remediation. Additionally, we reported real-life experience and challenges faced when trying to integrate auditing and compliance verification into OpenStack. We further conducted experiments to demonstrate the applicability of our approach. Our evaluation results show that formal methods can be successfully applied for large data centers with a reasonable overhead. However, this work does not support continuous compliance checking, which can enable monitoring various events, and triggering the verification process whenever a security property is affected by the changes. In the next chapter, we address this limitation through a runtime auditing mechanism for clouds.

Chapter 5
User-Level Runtime Security Auditing for the Cloud

In this chapter, we present an efficient user-level runtime security auditing framework in a multi-domain cloud environment. The multi-tenancy and ever-changing nature of clouds usually implies significant design and operational complexity, which may prepare the floor for misconfigurations and vulnerabilities leading to violations of security properties. Runtime security auditing may increase cloud tenants' trust in the service providers by providing assurance on the compliance with security properties mainly derived from the applicable laws, regulations, policies, and standards. Evidently, the Cloud Security Alliance has recently introduced the Security, Trust & Assurance Registry (STAR) for security assurance in clouds, which defines three levels of certifications (self-auditing, third-party auditing, and continuous, near real-time verification of security compliance) [19].

Motivating Example
Here, we provide a sketch of the gap between high-level standards and low-level input data, and the necessity of runtime security auditing.

- Section 13.2.1 of ISO 27017 [52], which provides security guidelines for the use of cloud computing, recommends *"checking that the user has authorization from the owner of the information system or service for the use of the information system or service…"*.
- The corresponding logging information is available in OpenStack [90] from at least three different sources:

  - Logs of user events (e.g., `router.create.end 1c73637 94305b c7e62 2899` meaning user 1c73637 from domain 94305b is creating a router).
  - Authorization policy files (e.g., `"create_router": "rule: regular_user"` meaning a user needs to be a regular user to create a router).
  - Database record (e.g., `1c73637 Member` meaning user 1c73637 holds the *Member* role).

© Springer Nature Switzerland AG 2019
S. Majumdar et al., *Cloud Security Auditing*, Advances in Information Security 76,
https://doi.org/10.1007/978-3-030-23128-6_5

- Continuously allocating and deprovisioning of resources and user roles for up to 100,000 users mean any verification results may only be valid for a short time. For instance, a re-verification might be necessary after certain frequently occurred operations such as: user create 1c73637 (meaning the 1c73637 user is created) and role grant member 1c73637 (meaning the member role is granted to the 1c73637 user).

Existing approaches can be roughly divided into three categories. First, the *retroactive* approaches (e.g., [63, 65]) catch security violations after the fact. Second, the *intercept-and-check* approaches (e.g., [13, 89]) verify security invariants for each user request before granting/denying it. Third, the *proactive* approaches (e.g., [13, 66, 89]) verify user requests in advance. Our work falls into the second category. Therefore, this work potentially prevents the limitation of the retroactive approaches and also requires no future change plan unlike proactive approaches (e.g., [13, 89]). In comparison with existing intercept-and-check solutions, our approach reduces the response time significantly and supports a wide range of user-level security properties.

Clearly, during the runtime security auditing, collecting and processing all the data again after each operation can be very costly and may represent a bottleneck for achieving the desired response time due to the performance overhead involved with data collection and processing operations (as reported in Sect. 5.4). In addition to data collection and processing, runtime verification of ever-changing clouds within a practical response time is essential and non-trivial. In this specific case, no automated tool exists yet in OpenStack for these purposes.

In the following, we provide an overview of a user-level runtime security auditing framework in a multi-domain cloud environment. First, we compile a set of security properties from both the existing literature on authorization and authentication and common cloud security standards. Second, we perform costly auditing operations (e.g., data collection and processing, and initial verification on the whole cloud) only once during the initialization phase so that later runtime operations can be performed in an incremental manner to reduce the cost of runtime verification significantly with a negligible delay. To this end, we rely on formal verification methods to enable automated reasoning and provide formal proofs or counterexamples of compliance. Third, we implement and integrate the proposed runtime auditing framework into OpenStack and report real-life experiences and challenges. Our framework supports several popular cloud access control and authentication mechanisms (e.g., role-based access control (RBAC) [33], attribute-based access control (ABAC) [47], and single sign-on (SSO)) with the provision of adding such more extensions. Finally, our experimental results confirm the scalability and efficiency of our approach.

5.1   User-Level Security Properties

We first show different attack scenarios based on authorization and authentication models. Then we formulate user-level threats as a list of security properties mostly derived from the cloud-specific standards, and finally discuss our threat model.

5.1.1   Models

We now describe RBAC, ABAC, and SSO models.

RBAC Model We focus on verifying multi-domain role-based access control (RBAC), which is adopted in real-world cloud platforms (as shown in Table 5.1). In particular, we assume the extended RBAC model as in [110], which adds multi-tenancy support in the cloud. The brief definitions of different components of this model are as follows. The details can be found in [110].

- Tenant[1]: A tenant is a collection of users who share a common access with specific privileges to the cloud instances.
- Domain: A domain is a collection of tenants, which draws an administrative boundary within a cloud.
- Object and operation: An object is a cloud resource, e.g., VM. An operation is an access method to an object. Object and operation together represent permissions.
- Token: A token is a package of necessary information used to authenticate prior to avail any operation.
- Group: Groups are formed for better user management.
- Trust: Trust is the concept, which enables delegation of duties over tenants or domains.
- Service: A service means a distributed cloud service.

Table 5.1 Support of different authorization and authentication mechanisms in major cloud platforms

Plugins	Cloud management platforms				
	OpenStack [90]	Amazon EC2 [3]	Microsoft Azure [71]	Google GCP [40]	VMware [113]
RBAC	•	•	•	•	•
ABAC	Blueprint [54]	•	Azure AD	Firebase	•
SSO	Federation	AWS directory	Microsoft account	G suite	myOneLogin

(•) indicates that the corresponding mechanism is fully supported in a cloud platform. Otherwise, the mechanism might be implemented using an existing service (as indicated) in the platform

---

[1]We interchangeably use the terms, tenant and project, in Fig. 5.1.

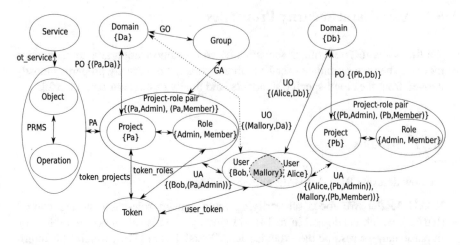

Fig. 5.1 Two domain instances of the access control model of [110] depicting the resultant state of the access control system after the exploit of the vulnerability, OSSN-0010. The shaded region and dotted arrows show an instance of the exploit described in Example 5.1

Example 5.1 Figure 5.1 depicts our running example, which is an instance of the access control model presented in [110]. In this scenario, Alice and Bob are the admins of domains, Da and Db, respectively, with no collaboration (trust) between the two domains; Pa and Pb are two tenants,[2] respectively, owned by the two domains. In such a scenario, we consider a real-world vulnerability, OSSN-0010,[3] found in OpenStack, which allows a tenant admin to become a cloud admin and acquire privileges to bypass the boundary protection between tenants, and illicitly utilize resources from other tenants while evading the billing. Suppose Bob has exploited this vulnerability to become a cloud admin. Figure 5.1 depicts the resultant state of the access control system after this attack. Therein, Mallory belonging to domain, Da, is assigned a tenant-role pair (Pb, Member), which is from domain, Db. This violates the security requirement of these domains as they do not trust each other.

ABAC Model ABAC [47], is considered as a strong candidate to replace RBAC in Sandhu [102], which identifies several limitations of RBAC and thus emphasizes the importance of ABAC specially for large infrastructures (e.g., cloud). In fact, major cloud platforms have started supporting ABAC (as shown in Table 5.1). We briefly review the central concepts of ABAC model here, while leaving more details to [47]. Attributes are pairs of names and values, and associated with different entities. The attribute is considered as a function, which takes users or objects as inputs and

---

[2]We interchangeably use the terms, tenant and project in Figs. 5.1 and 5.2.

[3]Keystone exposes privilege escalation vulnerability, available at: https://wiki.openstack.org/wiki/OSSN/OSSN-0010.

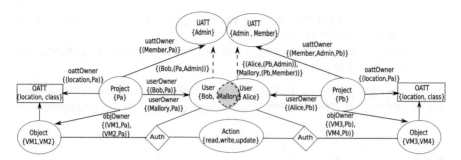

Fig. 5.2 Two tenant instances of the access control model of [99] depicting the resultant state of the access control system after the exploit of the vulnerability, OSSN-0010. The shaded region and dotted arrows show an instance of the exploit described in Example 5.2

returns a value from the attribute's scope. A user is a person or a non-person entity, e.g., an application, which requests for actions on objects. A user is described with a set of attributes (*UATT*), e.g., name, salary, role, etc. Objects are system resources (e.g., VMs, files, etc.) which can be accessed only by authorized users. Object attributes (*OATT*) represent resource properties such as risk level, classification, and location. Actions are the list of allowed operations on an object by a user. In this work, we mainly use two ABAC functions, i.e., user attribute (*UATT*) and object attribute (*OATT*).

Example 5.2 Figure 5.2 depicts our running example for ABAC and shows a similar attack scenario as Example 5.1. The model in the figure is an instance of the access control model presented in [99] and shows the resultant state of the access control system after this attack.

SSO Mechanism SSO, which is a popular cloud authentication extension and supported by major cloud platforms (shown in Table 5.1), only requires a single user action to permit a user to access all authorized computers and systems. In this work, we detail two SSO protocols: OpenID [83] and SAML [78] supported by OpenStack and many other cloud platforms.

However, there are several attacks (e.g., [5, 43, 79, 115]) on two above-mentioned SSO protocols. The following describes several security concerns specific to these protocols.

- In SAML, there is no communication between service provider (SP) and identity provider (IdP). Therefore, an SP maintains a list of trusted IdPs, and any ID generated by an IdP that is not in this list must be strictly restricted.
- On the other hand, OpenID accepts any IdP by communicating with the corresponding relying party (RP), which provides the login service as a third party. Therefore, a proper synchronization between IdP and RP is essential for the OpenID protocol. Otherwise, it may result following security critical incidents:

– Logging out from IdP may not ensure logging out from RP, and thus an unauthorized session is possible.
– Linking to an existing account with an OpenID without any authentication may result in unauthorized access.

To address such security concerns and to be compliant with aforementioned cloud-specific security standards, we devise security properties in the next subsection.

5.1.2   Security Properties

Table 5.2 presents an excerpt of the list of user-level security properties that we identify from the access control and authentication literature, relevant standards (e.g., ISO 27002 [51], NIST SP 800-53 [77], CCM [17], and ISO 27017 [52]), and the real-world cloud implementation (e.g., OpenStack). Even though some properties (e.g., cyclic inheritance) are not directly found in any standard, they are included based on their importance and impact described in the literature (e.g., [42]).

RBAC Security Properties RBAC-specific security properties are shown in Table 5.2. For our running example, we will focus on following two properties. Common ownership: based on the challenges of multi-domain cloud discussed in [42, 110], users must not hold any role from another domain. Minimum exposure: each domain in a cloud must limit the exposure of its information to other domains [110].

Example 5.3 The attack scenario in Example 5.1 violates the common ownership property. According to the property, Mallory must not hold a role member in tenant, Pb, belonging to domain, Db, because Mallory belongs to domain, Da, and there exists no collaboration between domains, Da and Db.

ABAC Security Properties Table 5.2 provides an excerpt of ABAC-related security properties supported by our auditing system. Some properties are specific to ABAC, and rests are adopted from RBAC. We only discuss the following properties, which are extended or added for ABAC.

• Consistent constraints: Jin et al. [55] define constraints for different basic changes in ABAC elements, e.g., adding/deleting users/objects. After each operation, certain changes are necessary to be properly applied. This property verifies whether all constraints have been performed.
• Common ownership: For ABAC, the common ownership property also includes objects and their attributes so that an object owner and the owner of the allowed user performing certain actions on that object must be the same.

Authentication-Related Security Properties Table 5.2 shows an excerpt of security properties related to generic authentication mechanisms and extensions (e.g., SSO). We discuss the SSO-related properties as follows. Brute force protection:

Table 5.2 An excerpt of user-level security properties

Properties	Standards				RBAC	ABAC	SSO
	ISO27002 [51]	ISO27017 [52]	NIST800 [77]	CCM [17]			
Role activation [53]	13.2.2b	15.2.2b	AC-1	IAM-09	•	•	
Permitted action [53]	11.2.1.b, 1.2.2c	13.2.1b, 13.2.2c	AC-14	IAM-10	•	•	
Common ownership [42]	11	13	AC	IAM	•	•	
Minimum exposure [110]	11.6.1	13.4.1	AC-4	IAM-04,06	•	•	
Separation of duties [110]	11.6.2	13.6.2	AC-5	IAM-02,05	•		
Cyclic inheritance [42]					•		
Privilege escalation [42]	11.2.2.b	13.2.2b	AC-6	IAM-08	•	•	
Cardinality [53]	11.2.4	13.2.4	AC-1		•	•	
Consistent constraints [55]							
Add/delete user	11.5.1	13.4.2	AC-7,9	IAM-02		•	
Modify user attributes	13.2.2b	15.2.2b	AC-1	IAM-09		•	
Add/delete object	13.2.2b	15.2.2b	AC-1	IAM-09		•	
Modify object attributes	13.2.2b	15.2.2b	AC-1	IAM-09		•	
Session de-activation [77]	11.5.5	13.2.8	AC-12		•	•	
User-access validation [52]	11.5.2	13.4	AC-3	IAM-10			•
User-access revocation [51]	11.2.1h	13.2.1h	AC-2	IAM-11			•
No duplicate ID [17]	11.5.2	13.5.2	AC-2	IAM-12			
Secure remote access [52]	11.4.2	13.4.2	AC-17	IAM-02,07			
Secure log-on [52]	11.5.1	13.4.2	AC-7,9	IAM-02			
Session time-out [77]	11.5.5	13.2.8	AC-12				•
Concurrent session [77]		13.5.4	AC-10				
Brute force protection	11.2.2.b	13.2.2b	AC-1	IAM-09			•
No caching	11.2.1.b, 1.2.2c	13.2.1b, 13.2.2c	AC-14	IAM-10			•

account lockout, CAPTCHA, or any such brute force protection must be applied in SSO. No caching: SSO and associated applications should set no-cache and no-stored-cache directives. User access revocation: logout from one application must end sessions of other applications. User access validation: only a valid authentication token must pass the authentication step.

5.1.3   Threat Model

Our threat model is based on two facts. First, our solution focuses on verifying the security properties specified by cloud tenants, instead of detecting specific attacks or vulnerabilities (which is the responsibility of IDSes or vulnerability scanners). Second, the correctness of our verification results depends on the correct input data extracted from logs and databases. Since an attack may or may not violate the security properties specified by the tenant, and logs or databases may potentially be tampered with by attackers, our results can only signal an attack in some cases. Specifically, the in-scope threats of our solution are attacks that violate the specified security properties and at the same time lead to logged events. The out-of-scope threats include attacks that do not violate the specified security properties, attacks not captured in the logs or databases, and attacks through which the attackers may remove or tamper with their own logged events. More specifically, in this work, we focus on user-level security threats and rely on existing solutions (e.g., [13, 66]) to identify virtual infrastructure level threats. We assume that, before our runtime approach is launched, an initial verification is performed and potential violations are resolved. However, if our solution is added from the commencement of a cloud, obviously no prior security verification (including the initial phase) is required. We also assume that tenant-defined policies are properly reflected in the policy files of the cloud platforms.

5.2   Runtime Security Auditing

This section presents our runtime security auditing framework for the user-level in the cloud.

5.2.1   Overview

Figure 5.3 shows an overview of our runtime auditing approach. This approach contains two major phases: (1) initialization, where we conduct a full verification on the collected and processed cloud data and (2) runtime, where we incrementally verify the compliance upon dynamic changes in the cloud. The initialization

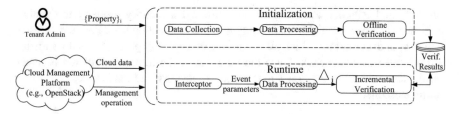

Fig. 5.3 An overview of the runtime security auditing approach

phase is performed only once through an offline verification. This phase performs all costly operations such as data collection and processing, and an initial full compliance verification for a list of security properties. The initial verification result is stored in the result repository. For the latter, we devise an incremental verification approach to minimize the workload at the runtime. During the runtime phase, each management operation (e.g., create/delete a user/tenant) is intercepted, its parameters are processed with the previous result, and finally the verification engine evaluates the compliance and provides the latest verification result. We elaborate major phases of our system as follows.

5.2.2   Initialization Phase

Our runtime auditing approach at first requires one-time data collection and processing, and full verification of the cloud, namely the initialization phase, against a list of security properties. Initially, we collect all necessary data from the cloud, which are in different formats and in different levels of abstractions. Therefore, we further process these data to convert the format and correlate them to identify required relationships for considered security properties. Then, we generate inputs for the verification engine incorporating the processed data in the previous step. Finally, the verification engine checks the compliance for a list of security properties and provides an initial verification result.

The collection engine is responsible for collecting the required data in a batch mode from the cloud management system. The role of the processing engine is to filter, format, aggregate, and correlate this data. The required data is distributed throughout the cloud and in different formats (e.g., files and databases). The processing engine must pre-process the data in order to provide specific information needed to verify given properties. A final processing step is to generate and store the inputs to be used by the compliance verification engine. Note that the format of the inputs depends on the selected back-end verification engine.

The compliance verification engine is to perform the actual verification of the security properties. We use formal methods to capture the system model and to verify properties, which facilitates automated reasoning and is generally more practical and effective than manual inspection. If a security property is violated, evidences can be obtained from the output of the verification back-end, e.g., a set of real data in the cloud for which all conditions of a security property are not satisfied are provided as a feedback. Once the outcome of the initial verification is ready, results and evidences are stored in the result repository and made accessible to the runtime engine.

5.2.3   Runtime Phase

The initialization phase conducts an offline verification, where we verify security properties on the whole cloud. However, verifying the whole cloud after each configuration change is very expensive. Alternatively, we intercept each event and verify the impact of the events in an incremental manner, where we perform runtime verification on a minimal dataset based on the current change to provide a better response time, to catch a security violation.

We update the verification results continuously by verifying the cloud at runtime. The runtime verification is event driven and property specific. Table 5.3 shows the events that may affect the result of verification for certain properties. The bottom part of Fig. 6.20 depicts the steps of this phase. We intercept each operation request generated from the cloud management interface. We further retrieve the parameters of the request and pass them to the data processing module. The data processing module performs similarly as described in Sect. 5.2.2. However, during the runtime phase, mostly partial data are sent for the incremental verification for each security property. Thus, the incremental verification is only conducted on the impact of the current change. Then, the final verification result is inferred from the result of current incremental verification and the previous result. Incremental verification is performed using any of the two methods: (1) *deltaVerify*, where compliance verification mechanism discussed in the initialization phase is applied on the delta data, and (2) *customAlgo*, where security property specific customized algorithms are performed. We discuss our runtime verification algorithm in more details in Sect. 5.3.2. In the following examples, we assume that the previous verification result for a specific property is stored in $Result_{t0}$, the parameters of the intercepted event is in Δ_i and the updated result will be stored in $Result_t$. For example, all user-role pairs violating the common ownership property at time $t0$ are stored in $Result_{t0}$ as $\{Mallory, Da, (Pb, Member), Db\}$.

Example 5.4 Table 5.3 shows that the verification result for the common ownership property may change for following events: grant role, delete role, delete user, delete tenant, and delete domain. Upon intercepting any of these events, we conduct incremental verification as follows:

Table 5.3 Events that influence verification results for certain properties

Properties	Create user	Create role	Create tenant	Create domain	Create operation	Create object	Delete user	Delete role	Delete tenant	Delete domain	Delete operation	Delete object	Grant role	Revoke token	Enable user	Enable role	Enable tenant	Enable domain	Disable user	Disable role	Disable tenant	Disable domain
Common ownership	•						•		•	•			•									
Permitted action							•	•	•	•			•									
Minimum exposure					•	•					•	•										
Role activation							•	•	•													
Separation of duties							•	•	•	•			•									
Privilege escalation		•					•	•	•	•			•									
Cardinality		•					•	•	•	•			•			•	•	•	•	•	•	•
Cyclic inheritance		•						•														
No duplicate ID	•	•	•	•	•	•	•	•	•	•	•	•	•									
User access validation								•						•						•		
User access revocation							•	•	•	•			•	•		•	•	•	•	•	•	•
Secure remote access	•													•	•							
Secure log-on	•													•	•							
Session time-out							•		•	•				•		•	•	•	•	•	•	•
Concurrent session							•		•					•					•	•	•	•

- Grant a role: Each role assignment alone may affect the common ownership property, and hence, it does not depend on the previous assignments. Therefore, upon a grant role request, we only verify the parameters of the intercepted event using the *deltaVerify* method and combine the obtained result with the previous result ($Result_{t0}$) to infer the updated result ($Result_t$).
- Delete a role: If the deleted role (Δ_i) is present in the previous result (i.e., $\Delta_i \in Result_{t0}$), then we update the current result by removing that role (i.e., $Result_t = Result_{t0} - \Delta_i$). Otherwise, the result remains unchanged (i.e., $Result_t = Result_{t0}$). Deleting a user/tenant/domain can be similarly handled.

Example 5.5 Similarly, upon intercepting any of the events marked for the permitted action property in Table 5.3, we conduct incremental verification as follows:

- Grant a role: If the granted role (Δ_i) is present in the previous result (i.e., $\Delta_i \in Result_{t0}$), then we update the current result by removing that role (i.e., $Result_t = Result_{t0} - \Delta_i$). Otherwise, the result remains unchanged (i.e., $Result_t = Result_{t0}$).
- Delete a role: If the deleted role (Δ_i) is present in the previous result (i.e., $\Delta_i \in Result_{t0}$), then we update the current result by removing that role (i.e., $Result_t = Result_{t0} - \Delta_i$). Otherwise, the result remains unchanged (i.e., $Result_t = Result_{t0}$). Deleting a user/tenant/domain can be similarly handled.

Identifying the Impacts of Events By observing the impacts of cloud events, Table 5.3 lists all events that may change the verification result of certain security properties. However, identifying the impacts of events in cloud can be challenging. Also, the completeness of the method of identifying the impacts relies on the specifications of APIs by the cloud platforms. In this work, we mainly follow two methods (i.e., API documentation and infrastructure change inspection) proposed by Bleikertz et al. [13]. Firstly, we go through the API documentation provided by cloud platforms to obtain API specifications including their functionality, parameters, and impacts on the infrastructure. Secondly, we perform different events and observe the infrastructure configuration change to capture the impact of those events. Finally, we combine this knowledge with the definition of security properties to populate Table 5.3.

Provision of Enriching the Security Property List Beside the security properties in Sect. 5.1.2, tenants might intend to add new security properties over time. Our framework provides the provision of adding new security properties through following simple steps. First, the new security property is defined in the cloud system context, which can be simply performed by following our existing techniques discussed in Sect. 5.1.2 to apply high-level standard terminologies to cloud-specific resources. Next, the property is translated to the first-order logic and then to Constraint Satisfaction Problem (CSP) constraints, and in many cases the existing relations discussed in Sect. 5.2.4 can be re-used as they include basic relations such as *belongs to, owner of, authorized for,* etc. Our data collection engine already collects data from all relevant sources of data of a cloud platform regardless of security properties. Therefore, no extra effort is needed in the data collection phase, unless the new property requires data from a different layer in the cloud (e.g., SDN). Then, the data processing effort for a new property mainly involves building a correlation between data from different sources, because other processing steps are mostly property independent. The remaining initial verification step is only to add constraints of the new property to the verification list. Finally, we identify the events that may alter the verification result of the new property by re-utilizing the knowledge of impacts of events and perform the runtime verification through incremental steps either using the deltaVerify method or by a customized algorithm (as in Sect. 5.3.2). Additionally, whenever there is any change in the event specification for a cloud system, we capture the update on impacts (if any) of events on the security properties.

5.2.4 Formalization of Security Properties

As a back-end verification mechanism, we formalize verification data and properties as Constraint Satisfaction Problem (CSP) and use a constraint solver, namely Sugar [108], to validate the compliance. CSP allows formulation of many complex problems in terms of variables defined over finite domains and constraints. Its generic goal is to find a vector of values (a.k.a. assignment) that satisfies all constraints expressed over the variables. If all constraints are satisfied, the solver returns SAT, otherwise, it returns UNSAT. In the case of a SAT result, a solution to the problem is provided. In our case, we formalize each security property in CSP to verify in Sugar. After verification, Sugar provides proper evidence (a.k.a counterexamples) of a violation (if any) of a security property. In the following, we first provide a generic description of model formalization, then illustrate examples of property formalization, and finally show some counterexamples for those security properties.

Model Formalization Referring to Figs. 5.1 and 5.2, entities are encoded as CSP variables with their domains definitions (over integer), where instances are values within the corresponding domain. For example, *User* is defined as a finite domain ranging over integer such that (domain $User$ 0 max_user) is a declaration of a domain of users, where the values are between 0 and max_user. Relationships and their instances are encoded as relation constraints and their supports, respectively. For example, *AuthorizedR* is encoded as a relation, with a support as follows: (relation $AuthorizedR$ 3 (supports (r1,u1,t1) (r2,u2,t2))). The support of this relation will be fetched and pre-processed in the data processing step. The CSP code mainly consists of four parts:

- *Variable and domain declaration.* We define different entities and their respective domains. For example, u and op are entities (or variables) defined, respectively, over the domains *User* and *Operation*, which range over integers.
- *Relation declaration.* We define relations over variables and provide their support from the verification data.

- *Constraint declaration.* We define the negation of each property in terms of predicates over the involved relations to obtain a counterexample in case of a violation.
- *Body.* We combine different predicates based on the properties to verify using Boolean operators.

Properties Formalization for RBAC Security properties are presented as predicates over relation constraints and predicates. We detail two representative properties in this work. We first express these properties in first-order logic [8] and then in their CSP formalization (using Sugar syntax). Table 5.4 summarizes the relations that we use in these properties.

1. Common ownership: Users are authorized for the roles that are only defined within their domains.

$$\forall u \in \text{User}, \forall d \in \text{Domain}, \forall r \in \text{Role}, \forall t \in \text{Tenant} \qquad (5.1)$$

$$\text{BelongsToD(u, d)} \wedge \text{AuthorizedR(u, t, r)} \longrightarrow$$

$$\text{TenantRoleDom(t, r, d)}$$

The corresponding CSP constraint is

$$(\text{and} \quad \text{BelongsToD(u, d) AuthorizedR(u, t, r)} \qquad (5.2)$$

Table 5.4 Correspondence between relations in our formalism and relationships/entities in Fig. 5.1

Relations in properties	Evaluate to *True* if	Corresponding relations in Fig. 5.1
AuthorizedOp(d, t, u, r, o, op)	In domain *d*, and tenant *t*, the user *u*, with the role *r* is authorized to perform operation *op* on object *o*	UA, PO, tenant-role pair, PA, PRMS
OwnerD(od, t, o)	Domain *od* is the owner of the object *o* in tenant *t*	PO, tenant-role pair, PA
AuthorizedR (u, t, r)	User *u* belonging to tenant *t* is authorized for the role *r*	UA, tenant-role pair
BelongsToD(u, d)	User *u* belongs to the domain *d*	UO
TenantRoleDom (t, r, d)	Role *r* is defined within the domain *d* in tenant *t*	PO, tenant-role pair
LogEntry(d, t, u, r, o, op)	Operation *op* on object *o* is actually performed by user *u* having role *r* in tenant *t* and domain *d*	ND
ActiveToken(tok, d, t, u, r, time)	Token *tok* is active at time *time* and in use by user *u* having role *r* in tenant *t* and domain *d*	UA, token_tenants, token_roles, PO, tenant-role pair

Note that one of the relations (in third column) is denoted by ND as it is inferred from dynamic data (e.g., logs)

$$(\text{not} \quad \texttt{TenantRoleDom(t, r, d)}))$$

2. Minimum exposure: We assume that the user access is revoked properly and that each domain's administrator may share a set of objects (resources) with other domains. The administrator defines accordingly a policy governing the shared objects, the allowed domains for a given object, and the allowed actions for a given domain with respect to a specific object. During data processing, we recover for each domain, the set of foreign objects (belonging to other domains) and the actual operations performed on those objects (from the logs). This property allows checking whether the collected and correlated data complies with the defined policy of each domain.

$$\forall \texttt{d, od} \in \texttt{Domain}, \forall \texttt{o} \in \texttt{Object}, \forall \texttt{op} \in \texttt{Operation}, \qquad (5.3)$$

$$\forall \texttt{r} \in \texttt{Role}, \forall \texttt{t} \in \texttt{Tenant}, \forall \texttt{u} \in \texttt{User}$$

$$\texttt{LogEntry(d, t, u, r, o, op)} \wedge \texttt{BelongsTo(u, d)} \wedge$$

$$\texttt{OwnerD(od, t, o)} \longrightarrow \texttt{AuthorizedOp(d, t, u, r, o, op)})$$

The CSP constraint for this property is:

$$(\texttt{and(and} \quad \texttt{LogEntry(d, t, u, r, o, op)} \qquad (5.4)$$

$$\texttt{OwnerD(od, t, o)} \texttt{ BelongsTo(u, d))}$$

$$(\texttt{not} \quad (\texttt{AuthorizedOp(d, t, u, r, o, op)})))$$

Properties Formalization for ABAC   In the following, we show formalization of one security property for ABAC.

Common ownership: Formally the common ownership property is violated in the following conditions: $userOwner(u) \neq uattOwner(userRole_i(u))$ OR $objOwner(o) \neq oattOwner(objRole_{i,j}(o))$. Through this extension, we complement the previous definition, and the property is now more general in the sense that we can identify the misconfiguration in defining policies for an object. Following example further explains this benefit. Alice is a user (from the user set, U) owned by the domain, $d1$. Alice holds a member role in the domain, $d2$, expressed as $userRole_2(Alice)$. The owner of this role is the domain, d2 (inferred from the $uattOwner(userRole2(Alice))$ relationship). This situation violates the common ownership property, as the first part of the condition (i.e., $userOwner(u) \neq uattOwner(userRole_i(u))$) is true. Additionally, there is an object i.e., VM1 (from the object set O) owned by the domain, $d2$. The policy related to $VM1$ states that a user with the *member* role of the $d2$ domain can *read* from $VM1$ (as described in $objRole1, 2(VM1)$). To verify the owner of the role that policy allows certain action on the object using the $oattOwner(objRole1, 2(VM1))$ relation. In this case, $objOwner(o) \neq oattOwner(objRole_{i,j}(o))$ is false; hence, the property is preserved.

Since ABAC is more expressive, there might be a larger set of properties for ABAC (as shown in Table 5.2). However, the verification complexity depends more on the security properties and less on the model. For example, the common ownership, permitted action, and minimum exposure properties show different level of complexities, as shown through their formal representation and as supported by the experiment results in Sect. 5.4.

Properties Formalization for SSO In the following, we present formalization steps of one SSO related security property (i.e., user access revocation). The user access revocation property is for the token-based user access. At a given time, for active tokens, we check that none of the situations leading to their revocation has been occurred. Function *TimeStamp(tok)* returns the token expiration time.

$$\forall tk \in \texttt{Token}, \forall r \in \texttt{Role}, \forall t \in \texttt{Tenant}, \qquad (5.5)$$

$$\forall u \in \texttt{User}, \forall d \in \texttt{Domain}$$

$$\texttt{ActiveToken(tk, d, t, u, r, Time)} \longrightarrow$$

$$\texttt{AuthorizedR(u, t, r)} \wedge \quad \texttt{IsActiveR(r, t, u)} \wedge$$

$$\texttt{BelongsToD(u, d)} \wedge \texttt{IsValidU(u)} \wedge \texttt{IsValidD(d)}$$

$$\wedge \texttt{IsvalidT(t)} \wedge \texttt{TimeStamp(tk)} > \texttt{Time}$$

Thus, the corresponding CSP constraint is:

$$\texttt{(and   ActiveToken(tk, d, t, u, r, time)(or   (not} \qquad (5.6)$$

$$\texttt{(not   AuthorizedR(u, t, r))(not   IsActiveR(r, t, u))}$$

$$\texttt{(IsValidU(u))(notIsvalidT(t))(notBelongsToD(u, d))}$$

$$\texttt{(notIsValidD(d))(not(> TimeStamp(tk)Time))))}$$

Evidences of Violations Our auditing system using the formal verification tool, Sugar, individually identifies the causes (a.k.a. counterexamples) for each security property violated in the cloud. With the following examples, we show how our system can locate the cause of the violations.

Example 5.6 The CSP predicate for the common ownership property is as follows: (*and BelongsToD(u, d) AuthorizedR(u, t, r) (not Tenant-RoleDom(t, r, d))*). The property is violated when a user from one domain holds a tenant-role pair from another domain. In other words, in case of a violation there exists at least a set of predicates as follows: (*and BelongsToD(u1, d1) AuthorizedR(u1, t2, r2) (notTenant-RoleDom(t2, r2, d1))*); meaning that the user, *u1*, from domain, *d1*, holds a role pair, *t2-r2*, which is not from domain, *d1*. In such cases, our auditing system using Sugar identifies that the (u1, d1, t2, r2) tuple is the cause for a violation of the common ownership property. In Sect. 5.3, Example 5.7 further extends this example to show concrete examples of evidences provided by our auditing system.

5.3 Implementation

In this section, we first illustrate the architecture of our system. We then detail our auditing framework implementation and its integration into OpenStack along with the challenges that we face and overcome.

5.3.1 Architecture

Figure 5.4 shows a high-level architecture of our runtime verification framework. It has three main components: data collection and processing module, compliance verification module, and dashboard and reporting module. In the following, we describe different engines inside the data collection and processing module. The security property extractor identifies the sources of required data for a list of security properties. The event interceptor intercepts each management operation requested by the user in the cloud infrastructure system. The data collection engine interacts mainly with the cloud management system, the cloud infrastructure system (e.g., OpenStack), and elements in the data center infrastructure to collect various types of audit data. Then the data processing engine aids to build the correlation and to uniform the collected data. Our compliance verification module is responsible for the offline and runtime verification using the formal verification and validation (V&V) tools and our custom algorithms. Finally, the dashboard and reporting module interacts with the cloud tenant through the dashboard to obtain the tenant requirements and to provide the tenant with the verification results in a report. Tenant requirements encompass both general and tenant-specific security policies, applicable standards, as well as verification queries.

5.3.2 Integration into OpenStack

We focus mainly on three components in our implementation: the data collection and processing module, the compliance verification module, and dashboard and reporting module. In the following, we first provide background on OpenStack, and then describe our implementation details.

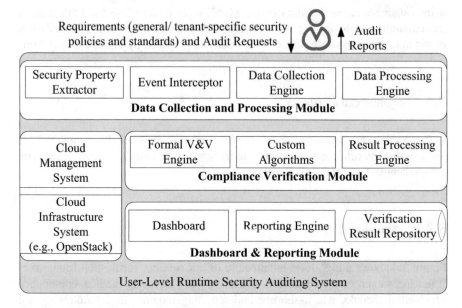

Fig. 5.4 A high-level architecture of our runtime verification framework

Background OpenStack [90] is an open-source cloud infrastructure management platform in which Keystone is its identity service, Neutron is its network component, Nova is its compute component, and Ceilometer is its telemetry.

Data Collection Engine The collection engine involves several components of OpenStack, e.g., Keystone and Neutron for collecting data from log files, policy files, different OpenStack databases, and configuration files from the OpenStack ecosystem to fully capture the configuration. We present hereafter different sources of data in OpenStack along with the current support for auditing offered by OpenStack. The main sources of data in OpenStack are logs, configuration files, and databases. Table 5.5 shows some sample data sources. The OpenStack logs are maintained separately for each service, e.g., Neutron, Keystone, in a directory named *var/log/component_name*, e.g., *keystone.log* and *keystone_access.log* are stored in the *var/log/keystone* directory. Two major configuration files, namely *policy.json* and *policy.v3cloudsample.json*, contain policy rules defined by both the cloud provider and tenant admins, and are stored in the *keystone/etc/* directory. The third source of data is a collection of databases, hosted in a MySQL server, that can be read using component-specific APIs such as Keystone and Neutron APIs. With the proper configuration of the OpenStack middleware, notifications for specific events in Keystone, Neutron, and Nova can be gathered from the Ceilometer database.

The effectiveness of a verification solution critically depends on properly collected evidences. Therefore, to be comprehensive in our data collection process,

Table 5.5 Sample data sources in OpenStack for relations in Table 5.4

Relations	Sources of data
AuthorizedOp	User, assignment, role in Keystone database and *policy.json* and *policy.v3cloudsample.json*
OwnerD	User, assignment in Keystone database and *policy.json*
AuthorizedR	User, tenant, assignment in Keystone database
BelongsToD	User, domain tables in Keystone database
TenantRoleDom	Tenant, assignment, domain tables in Keystone database
LoggedEntry	*keystone_access.log* and Ceilometer database
ActiveToken	Keystone database and *keystone_access.log*

we firstly check fields of all varieties of log files available in Keystone and more generally in OpenStack, all configuration files, and all Keystone database tables (18 tables). Through this process, we identify all possible types of data with their sources. Due to the diverse sources of data, there exist inconsistencies in formats of data. On the other hand, to facilitate verification, presenting data in a uniform manner is very important. Therefore, we facilitate proper formatting within our data processing engine.

Data Processing Engine Our data processing engine, which is implemented in Python, mainly retrieves necessary information from the collected data, converts it into appropriate formats, recovers correlation, and finally generates the source code for Sugar. First, our tool fetches the necessary data fields from the collected data, e.g., identifiers, API calls, timestamps. Similarly, it fetches access control rules, which contain API and role names, from *policy.json* and *policy.v3cloudsample.json* files. In the next step, our processing engine formats each group of data as an n-tuple, i.e., (user, tenant, role, etc.). To facilitate verification, we additionally correlate different data fields. In the final step, the n-tuples are used to generate the portion of the Sugar's source code, and the relationships for security properties (discussed in Sect. 5.2.4) are also appended with the code. Different scripts are needed to generate the Sugar source code for the verification of different properties, since relationships are usually property specific.

The logs generated by each component of OpenStack usually lack correlation. Even though Keystone processes authentication and authorization steps prior to a service access, Keystone does not reveal any correlated data. Therefore, we build the data correlation support within the processing engine. For example, we infer the relation (*user operation*) from the available relations (*user role*) and (*role operation*). In our settings, we have 61,031 entries in the (*user role*) relations for 60,000 users. The number of entries is larger than the number of users, because there are some users with multiple roles. With the increasing number of users having multiple roles, the size of this relation grows, and as a result, it increases the complexity of the correlation step.

Initial Compliance Verification The compliance verification module contains two major modules responsible for the initial verification and runtime verification, respectively. The pre-requisite formalization steps of the initial verification are already discussed in Sect. 5.2.4. Here, we explain different parts of a Sugar source code through a simple example and verification algorithm (as in Algorithm 1) in the following.

Listing 5.1 Sugar source code for the common ownership property

```
 1  //Declaration
 2  (domain Domain 0 500) (domain Tenant 0 1000)
 3  (domain Role 0 1000) (domain User 0 60000)
 4  (int D Domain) (int R Role)
 5  (int P Tenant) (int U User)
 6  //Relations Declarations and Audit Data as their Support
 7  (relation BelongsToD 2 (supports (100 401) (40569 123)
 8  (102 452) (145 404) (156 487) (128 463)))
 9  (relation AuthorizedR 3 (supports (100 301 225)
10  (40569 1233 9) (102 399 230) (101 399 231)))
11  (relation TenantRoleDom 3 (supports (301 225 401)
12  (1233 9 335) (399 230 452) (399 231 452)))
13  //Security Property: Common Ownership
14  (predicate(ownership D R U P)
15  (and (AuthorizedR U P R ) (BelongsToD U D)
16  (not(TenantRoleDom P R D)) ))
17  (ownership D R U P)
```

Example 5.7 Listing 5.1 is the CSP code to verify the common ownership property. Each domain and variable are first declared (lines 2–5). Then, the set of involved relations, namely *BelongsToD*, *AuthorizedR*, and *TenantRoleDom*, are defined and populated with their supporting tuples (lines 7–12), where the support is generated from actual data in the cloud. Then, the common ownership property is declared as a predicate, denoted by *ownership*, over these relations (lines 14–16). Finally, the predicate is instantiated (line 17) to be verified. As we are formalizing the negation of the properties, we are expecting the UNSAT result, which means that all constraints are not satisfied (i.e., no violation of the property). Note that the predicate is unfolded internally by the Sugar engine for all possible values of the variables, which allows to verify each instance of the problem among possible values of domains, users, and roles.

In this example, we also describe how a violation of the common ownership property may be caught by our verification process. Firstly, our program collects data from different tables in the Keystone database including *users*, *assignments*, and *roles*. Then, the processing engine converts the collected data and represents as tuples; for our example: (40569 123) (40569 1233 9) (1233 9 335), where Mallory: 405,69, Da: 123, Pb: 1233, member: 9 and Db: 335. Additionally, the processing engine interprets the property and generates the Sugar source code (as Listing 5.1) using processed data and translated property. Finally, the Sugar engine is used to verify the security properties. The CSP predicate for the common ownership property is as follows: $(and \ BelongsToD(u, d) \ AuthorizedR(u, t, r) \ (not \ Tenant - RoleDom(t, r, d)))$. As Mallory belongs to domain, Da, $BelongsToD(Mallory, Da)$ evaluates to true. Mallory has been authorized a tenant-role pair, $(Pb, member)$, thus $AuthorizedR(Mallory, Pb, member)$ evaluates to true. However, $TenantRoleDom(Pb, member, Da)$ evaluates to false, as the pair $(Pb, member)$ does not belong to domain Da. Then, the whole *ownership* predicate unfolded for this case is evaluated to true. In this case, the output of sugar is SAT, which confirms that Mallory violates the common ownership property and further presents the cause of the violation, i.e., $(d = 123, r = 9, t = 1233, u = 40569)$.

Runtime Verification Our runtime verification engine implements Algorithm 1. Firstly, the interceptor module intercepts each management operation based on the existing intercepting methods (e.g., audit middleware [91]) supported in OpenStack. Events are primarily created via the notification system in OpenStack; Nova, Neutron, etc. emit notifications in a JSON format. Here, we leverage the audit middleware in Keystone to intercept Keystone, Neutron, and Nova events by enabling the audit middleware and configuring filters. Secondly, the data processing engine handles the intercepted parameters to perform similar data processing operations as discussed previously. The processed data is denoted as Δ_i. Finally, the runtime verification engine performs incremental steps either using the deltaVerify method, which involves Sugar, or custom algorithms. Figures 5.5 and 5.6 show

---

Algorithm 1 Runtime compliance verification

1: **procedure** INITIALIZE(Properties,CloudOS)
2:     rawData = collectData(CloudOS)
3:     verData = processData(rawData)
4:     **for** each property $p_i \in Properties$ **do**
5:         $Result_{t0, pi}$ = Verify (p_i,verData)
6:     **end for**
7: **end procedure**
8: **procedure** RUNTIME(Event,$Result_{t0}$,Properties)
9:     **for** each property $p_i \in Properties$ **do**
10:         Δ_i = processData(event.parameters)
11:         **if** incremental-method(p_i) = $custom$ **then**
12:             custom-algo($event,p_i,Result_{t0, pi},\Delta_i$)
13:         **else**
14:             deltaVerify($event,p_i,Result_{t0, pi},\Delta_i$)
15:         **end if**
16:     **end for**
17:     return $Result_t$
18: **end procedure**
19: **procedure** DELTAVERIFY($event,p_i,Result_{t0, pi},\Delta_i$)
20:     $Result_{t, pi}$ = verify(p_i,Δ_i)
21: **end procedure**

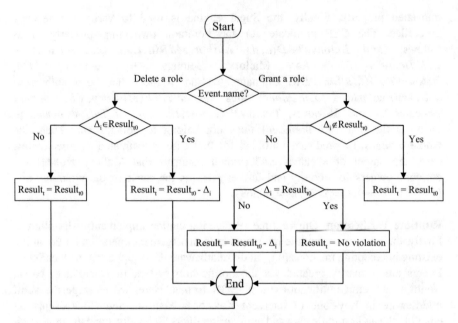

Fig. 5.5 Showing the runtime steps for the common ownership property

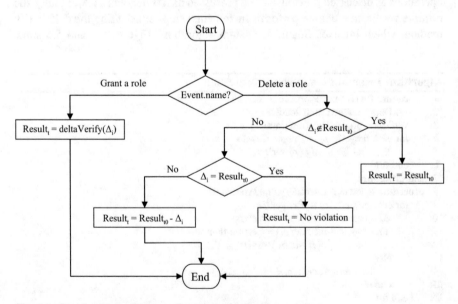

Fig. 5.6 Showing the runtime steps for the permitted action property

the incremental steps for the common ownership and permitted action properties, respectively.

There exist difficulties in locating relevant information, e.g., the initiator of Keystone API calls is missing, and in obtaining adequate notifications from Ceilometer for Keystone events. Therefore, to obtain sufficient and proper information about user events to conduct the auditing, we collect Neutron notifications from the Ceilometer database.

Dashboard and Reporting Module We further implement the web interface (i.e., dashboard) in PHP to place audit requests and view audit reports. In the dashboard, tenant admins can initially select different standards (e.g., ISO 27017, CCM V3.0.1, NIST 800-53, etc.). Afterwards, security properties under the selected standards can be chosen. Additionally, admins can select any of the following verification options: (1) runtime verification and (2) retroactive verification. Once the verification request is processed, the summarized verification results are shown and continuously updated in the verification report page. The details of any violation with a list of evidences are also provided. Moreover, our reporting engine archives all the verification reports for a certain period.

5.3.3 Integration to OpenStack Congress

To demonstrate the service agnostic nature of our framework, we further integrate our auditing method with OpenStack Congress [89]. Congress implements policy as a service in OpenStack in order to provide governance and compliance for dynamic infrastructure. Congress can integrate third-party verification tools using a data source driver mechanism. Using Congress policy language that is based on Datalog, we define several tenant-specific security policies as same as security properties described in Sect. 5.1.2. We then use our processed data to detect those security properties for multiple tenants. The outputs of the data processing engine in both cases of initialization and runtime are in turn provided as inputs for Congress to be asserted by the policy engine. This integrates compliance status for some policies whose verification is not yet supported by Congress (e.g., permitted action and minimum exposure).

5.4 Experiments

This section evaluates the performance of this work by measuring the execution time, and memory and CPU consumption.

5.4.1   Experimental Setting

We collect data from the OpenStack setup inside a lab environment. Our OpenStack version is Mitaka (2016.10.15) with Keystone API version v3. There are one controller node and three compute nodes, each having Intel i7 dual core CPU and 2 GB memory with the Ubuntu 16.04 server. To make our experiments more realistic, we follow recently reported statistics (e.g., [92] and [37]) to prepare our largest dataset consisting 100,000 users, 10,000 tenants, and 500 domains. For verification, we use the V&V tool, Sugar V2.2.1 [108]. We conduct the experiment for 12 different datasets in total. All data processing and V&V experiments are conducted on a PC with 3.40 GHz Intel Core i7 Quad core CPU and 16 GB memory, and we repeat each experiment 1000 times.

5.4.2   Results

The objective of the first set of our experiments (see Figs. 5.7 and 5.8) is to demonstrate the time and memory efficiency of our solution, and to compare the performance with a retroactive auditing approach similar as in [65]. Firstly, Fig. 5.7 shows time in milliseconds required for our runtime verification framework for the common ownership property. Our runtime verification requires a relatively expensive (i.e., about 2.5 s) initialization phase, similar to that of the retroactive approach. Afterwards, our runtime approach takes less than 100 ms; whereas, the retroactive approach always takes 2.5 s. Secondly, Fig. 5.8 compares time in

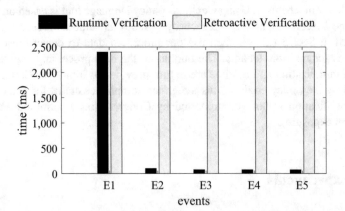

Fig. 5.7 Comparing the verification time required after each event for our system and the retroactive approach (e.g., [65]) for the common ownership property. Here, E1=initialization, E2=grant a role, E3=delete a role, E4=delete a user and E5=delete a tenant. The results are for our largest dataset

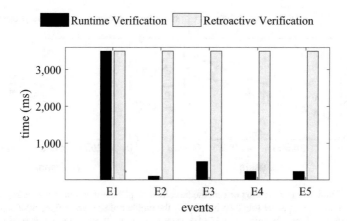

Fig. 5.8 Comparing the verification time required after each event for our system and the retroactive approach (e.g., [65]) for the permitted action property. Here, E1=initialization, E2=grant a role, E3=delete a role, E4=delete a user and E5=delete a tenant. The results are for our largest dataset

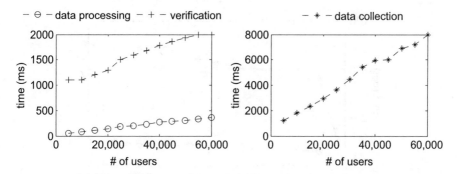

Fig. 5.9 Time required for each step during the initialization phase for the common ownership property while varying the number of users. Time for the data collection (right) is shown separately, as it is a one-time effort. In all cases, the number of domains is 500 and number of tenants is 10,000

milliseconds required for verifying the permitted action property by our framework and a retroactive verification method. For this property, we obtain results of the same nature as the previous one, i.e., requiring only a relatively expensive (i.e., about 3.5 s) initialization phase followed by runtime verification costing maximum 500 ms. For the permitted action property, after the *delete a role* event, a search for a certain role is performed; hence the verification time reaches the maximum value. Otherwise, verification time is within 100 ms for both properties.

Our second set of experiments (see Figures 5.9, 5.10, 5.11, 5.12, 5.13 and 5.14) is to demonstrate the time efficiency of individual phases of our solution. Firstly, Fig. 5.9 shows time in milliseconds required for data collection, data processing, and compliance verification during the initialization phase to verify the common

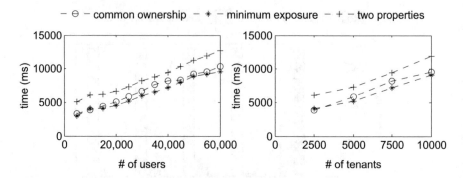

Fig. 5.10 Total time required to perform the initialization phase for common ownership, minimum exposure and both properties together, by varying the number of users with fixed 5000 tenants (left) and the number of tenants with fixed 30,000 users (right). In all cases, the number of domains is 500. Note that time in curves encompasses all three steps (collection, processing and verification). For the curve of two properties, data collection is performed one time

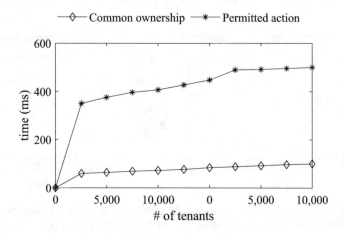

Fig. 5.11 Total time required to perform the runtime phase of the common ownership and permitted action properties, by varying the number of tenants with fixed 30,000 users. In all cases, number of domains is 500

ownership property for different cloud sizes (e.g., the number of users). The obtained results show that the verification execution time is less than 2 s for fairly large number of users. Knowing that this task is performed only once upon each request, we believe that this is an acceptable overhead for verifying a large setup. Figure 5.10 shows the total time required for separately performing the initialization phase for common ownership and minimum exposure properties, and also for both of the properties together. We can easily observe that the execution time is not a linear function of the number of security properties to be verified. In fact, we can see that verifying more security properties would not lead to a significant increase in the execution time. Figures 5.11 and 5.12 show the total time required

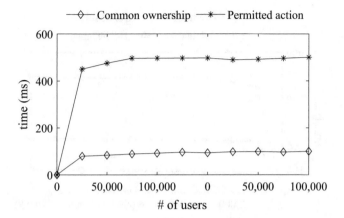

Fig. 5.12 Total time required to perform the runtime phase of the common ownership and permitted action properties, by varying the number of users with fixed 5000 tenants. In all cases, number of domains is 500

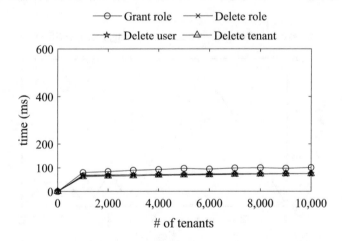

Fig. 5.13 Time required to perform the runtime phase of the common ownership property for different events, by varying the number of tenants with 10 users per tenant. In all cases, number of domains is 500

for separately performing the runtime phase for common ownership and permitted action properties for different cloud sizes. The obtained results support that the verification time for the permitted action (i.e., up to 500 ms) is more than that of the common ownership (i.e., up to 100 ms). Figures 5.13 and 5.14 further depict the effects of different events on the runtime phase for different security properties, while varying the number of tenants up to 10,000. As our runtime phase is an incremental approach and verifies mainly parameters of the events, the size of the cloud affects the verification time very less.

Our third experiment (see Figs. 5.15, 5.17, and 5.19) measures the CPU usage (in %) during the initialization and runtime phases. Figure 5.15 depicts the fact

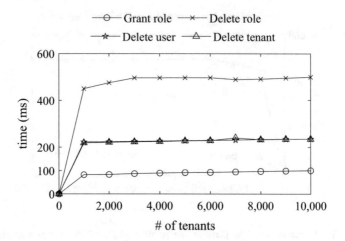

Fig. 5.14 Time required to perform the runtime phase of the permitted action property for different events, by varying the number of tenants with 10 users per tenant. In all cases, number of domains is 500

0-7s: data collection, 7-9s: data processing, 9-12s: verification

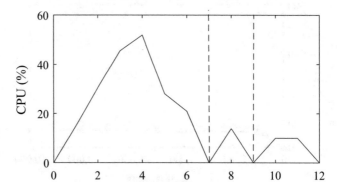

Fig. 5.15 CPU usage for each step during the initialization phase over time with 60,000 users, 10,000 tenants and 500 domains for the common ownership property

that the data collection step requires significantly higher CPU usage than the other two steps. However, the average CPU usage for data collection is 30%, which is reasonable since the verification process lasts only a few seconds. Note that, we conduct our experiment in a single PC; if the security properties can be verified through concurrent independent Sugar executions, we can easily parallelize this task by running several instances of Sugar on different VMs in the cloud environment. Thus, performing verification using the cloud or even with multiple servers possibly reduces the cost significantly. For the other two steps, the CPU cost is around 15%. In Fig. 5.17, we measure the peak CPU usage (in %) consumed by different steps

0-7s: data collection, 7-9s: data processing, 9-12s: verification

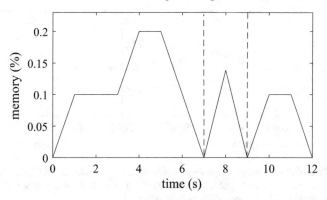

Fig. 5.16 Memory usage for each step during the initialization phase over time with 60,000 users, 10,000 tenants and 500 domains for the common ownership property

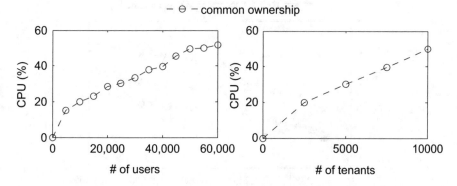

Fig. 5.17 Peak CPU usage to perform the initialization phase for the common ownership property by varying the number of users with 10,000 tenants (left) and number of tenants with 60,000 users (right). In both cases, there are 500 domains

while verifying the common ownership property. Accordingly, the CPU usage grows almost linearly with the number of users and tenants. We observe a significant reduction in the increase rate of CPU usage for datasets with 45,000 users or more. Note that, other properties show the same trend in CPU consumption, as the CPU cost is mainly influenced by the data collection step. Figure 5.19 shows that runtime phase expectedly requires negligible CPU (i.e., up to 4.7%) in comparison to the initialization phase (Fig. 5.17).

Our final experiment (Figs. 5.16, 5.18 and 5.20) measures the memory usage during the initialization and runtime phases. Figure 5.16 shows that the data collection step is the most costly in terms of memory usage. However, the highest memory usage observed during this experiment is only 0.2%. Figure 5.18 shows that the rise in memory consumption is only observed beyond 50,000 users (left) and 8000 tenants (right). We investigated the peak in the memory usage for 50,000

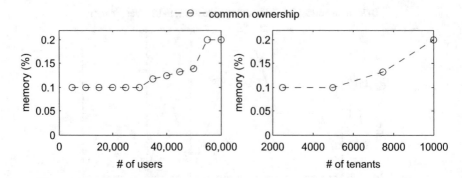

Fig. 5.18 Peak memory usage to perform the initialization phase for the common ownership property by varying the number of users with 10,000 tenants (left) and number of tenants with 60,000 users (right). In both cases, there are 500 domains

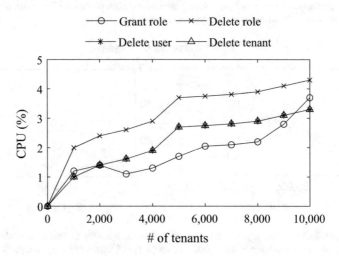

Fig. 5.19 Peak CPU usage to perform the runtime phase of the common ownership property for different events, by varying the number of tenants with 10 users per tenant. In all cases, the number of domains is 500

users and it seems that this is due to the internal memory consumption by Sugar. Figure 5.20 depicts the memory usage by our runtime phase and further supports that the runtime phase deals with significantly smaller dataset.

Although we report results for a limited set of security properties, the use of formal methods for verifying these properties shows very promising results. Particularly, we show that the time required for our solution grows very slowly with the number of security properties. As seen in Fig. 5.10, an additional security property adds only about 3 s to the initial effort. Therefore, we anticipate that verifying a large list of security properties would still be practical.

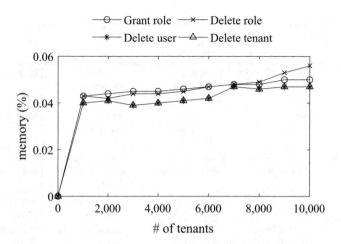

Fig. 5.20 Peak memory usage to perform the runtime phase of the common ownership property for different events, by varying the number of tenants with 10 users per tenant. In all cases, the number of domains is 500

5.5 Discussion

Adapting to Other Cloud Platforms Our solution is designed to work with most popular cloud platforms (e.g., OpenStack [90], Amazon EC2 [3], Google GCP [40], Microsoft Azure [71]) with a minimal one-time effort. Once a mapping of the APIs from these platforms to the generic event types is provided, the rest of the steps in our auditing system are platform-agnostic. Table 5.6 enlists some examples of such mappings.

Handling Extreme Situations There might be some extreme cases where our solution may act differently. For instance, if the cloud logging system fails resulting from any disruption or failure in the cloud, then our auditing system will be affected. As in our threat model (in Sect. 5.1.3), we assume that our solution relies on the correctness of the input data (including the logs) from the cloud. Any other failure or disruption in the cloud must be detected by our system. Also, if our system including the formal verification tool (e.g., Sugar) fails, till now there is no self-healing or self-recovery feature. Therefore, in this extreme case, the efficiency of the system will be affected and a full (instead of incremental) verification will be required to recover from this failure.

The Rationale Behind Our Incremental Approach The incremental verification of a given security property involves instantiating and solving the security property predicates for the affected elements in the supports of the involved relations (as stated in Sect. 5.2.4). Therefore, any modification to the system data resulted from cloud events (e.g., grant role, delete role, etc.) would not directly change the security

Table 5.6 Mapping event APIs of different cloud platforms to generic event types

Generic event type	OpenStack [90]	Amazon EC2-VPC [3]	Google GCP [40]	Microsoft Azure [71]
Create user	POST /v3/users	aws iam create-user	gcloud beta compute users create	az ad user create
Delete user	DELETE /v3/users/{user_id}	aws iam delete-user –user-name	gcloud beta compute users delete	az ad user delete
Assign role	/v3/users/{user_id} /roles/{role_id}	aws iam attach-role-policy	gcloud projects add-iam-policy-binding	az role assignment create
Create role	POST /v3/roles	aws iam create-role	gcloud beta iam roles create	az role definition create
Delete role	DELETE /v3/roles/{role_id}	aws iam delete-role	gcloud beta iam roles delete	az role definition delete

property expression itself although the corresponding support may need to be changed. For example, if a role is granted, the only change is that the relationships involving the entity role in the model would include a new element in their supports.

5.6   Conclusion

Despite existing efforts, runtime security auditing in cloud still faces many challenges. In this work, we proposed a runtime security auditing framework for the cloud with special focus on the user-level including different access control and authentication mechanisms e.g., RBAC, ABAC, and SSO, and we implemented and evaluated the framework based on OpenStack, a popular cloud management system. Our experimental results showed that our incremental approach in runtime verification reduces the response time to a practical level (e.g., less than 500 ms to verify 100,000 users). This response time is satisfactory when the management operations are manually done by the administrators. The current approach would be insufficient to provide the same response time in the case of batch execution for management operations, when these operations are executed in short intervals and if the subsequent operations impact the same property. Also, verifying sequence of events, in addition to our current method of verifying single event, may further reduce the impact of this concern. In the next chapter, we discuss a proactive security auditing approach, which overcomes the above-mentioned concerns and provides a practical response time.

Chapter 6
Proactive Security Auditing in Clouds

In this chapter, we present an automated learning-based proactive auditing system, namely *LeaPS* , which automatically learns probabilistic dependencies, and hence, addresses the inefficiencies of existing solutions. To this end, we describe a log processor, which processes (as discussed later) real-world cloud logs and prepares them for different learning techniques (e.g., Bayesian network and sequence pattern mining) to allow capturing dependency relationships. Unlike most learning-based security solutions, since we are not relying on learning techniques to detect abnormal behaviors, we avoid the well-known limitations of high false positive rates; any inaccuracy in the learning phase would only affect the efficiency, as will be demonstrated through experiments later in this chapter. We believe this idea of leveraging learning for efficiency, instead of security, may be adapted to benefit other security solutions. As demonstrated by our implementation and experimental results, LeaPS provides an automated, efficient, and scalable solution for different cloud platforms to increase their transparency and accountability to tenants.

As mentioned earlier in the book, security threats such as isolation breach in multi-tenant clouds cause persistent fear among tenants while adopting clouds [100]. To this end, security auditing in clouds can possibly ensure the accountability and transparency of a cloud provider to its tenants. However, the traditional approach of auditing, a.k.a. *retroactive auditing*, becomes ineffective with the unique nature (e.g., dynamics and elasticity) of clouds, which means the configurations of a cloud are frequently changed, and hence, invalidates the auditing results. To address this limitation and offer continuous auditing, the *intercept-and-check* approach verifies each cloud event at runtime. However, the sheer size of the cloud (e.g., 1000 tenants and 100,000 users in a decent-sized cloud [92]) can usually render the *intercept-and-check* approach expensive and non-scalable (e.g., over 4 min for a mid-sized cloud [13]). Since the number of critical events (i.e., events that may potentially breach security properties) to verify usually grows with the number of security properties supported by an auditing system, auditing larger clouds could incur prohibitive costs.

© Springer Nature Switzerland AG 2019   103
S. Majumdar et al., *Cloud Security Auditing*, Advances in Information Security 76,
https://doi.org/10.1007/978-3-030-23128-6_6

To this end, the proactive approach (e.g., [66]) is a promising solution and specifically designed to ensure a practical response time. Such an approach prepares for critical events in advance based on the so-called dependency models that indicate which events lead to the critical events [66, 123]. However, a key limitation of existing proactive approaches (including our previous work [66]) is that their dependency models are typically established through manual efforts based on expert knowledge or user experiences, which can be error-prone and tedious especially for large clouds. Moreover, existing dependency models are typically static in nature in the sense that the captured dependencies do not reflect runtime dynamic event patterns. A possible solution is to automatically learn probabilistic dependencies from the historical data (e.g., cloud logs). However, the log formats in current cloud platforms (especially, in OpenStack [90], which is one of the most popular cloud management platforms) are unstructured and not ready to be fed into different learning tools. Furthermore, due to the diverse formats of logs in different versions of the cloud platform, the log processing task becomes more difficult. Therefore, to enable log analysis (e.g., learning dependency models for proactive auditing), the need of a log processing approach addressing different real-world challenges (which are discussed in Sect. 6.2.2) and preparing raw logs for different learning tools is evident.

To address those limitations, our key idea is to design a log processor, which prepares the inputs for different learning techniques, to learn probabilistic (instead of deterministic) dependencies, and to automatically extract such a model from processed logs. Specifically, we first conduct case studies on cloud log formats in different OpenStack deployments including a real community cloud and enumerate all challenges related to raw log processing to automate different learning mechanisms. Second, we design a log processor that addresses all challenges identified in our investigation, and provides inputs for different learning techniques (e.g., Bayesian network and sequence pattern mining). Third, we propose a new approach to automatically generate the probabilistic dependency models from the processed logs. Fourth, we provide detailed methodology and algorithms for our learning-based proactive security auditing system, namely *LeaPS* , including the log processor, learning component, and proactive verification component. We describe our implementation of the proposed system based on OpenStack [90] and demonstrate how the system may be ported to other cloud platforms (e.g., Amazon EC2 [3] and Google GCP [40]). Finally, we evaluate our solution through extensive experiments with both synthetic and real data. The results confirm our solution can achieve practical response time (e.g., 6 ms to audit a cloud of 100,000 VMs) and significant improvement over existing proactive approaches (e.g., about 50% faster), and our log processor can be adopted by different learning techniques efficiently (e.g., only 18 ms to execute different sequence pattern mining algorithms for 50,000 events).

6.1 Overview

In this section, we present a motivating example, describe the threat model, and provide an overview of our proposed solution.

6.1.1 Motivating Example

The upper part of Fig. 6.1 depicts several sequences of events in a cloud (from Session N to Session $N + M$). The critical events, which can potentially breach some security properties, are shown shaded (e.g., $E2$, $E5$, and $E7$). The lower part of the figure illustrates two different auditing approaches of such events. We discuss their limitations below to motivate our solution.

- With a traditional runtime verification approach, most of the verification effort (depicted as boxes filled with vertical lines) is performed after the occurrence of the critical events, while holding these related operations blocked until a decision is made; consequently, such solutions may cause significant delays in operations.
- In contrast, a proactive solution will pre-compute most of the expensive verification tasks well ahead of the critical events in order to minimize the response time. However, this means such a solution would need to first identify patterns of event dependencies, e.g., $E1$ may lead to a critical event ($E2$), such that it may pre-compute as soon as $E1$ happens.
- Manually identifying patterns of event dependencies for a large cloud is likely expensive and non-scalable. Indeed, a typical cloud platform allows more than

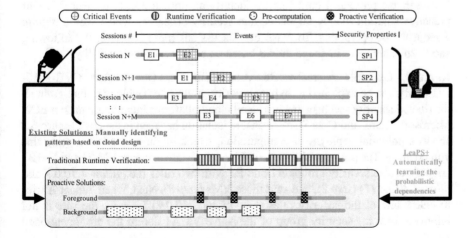

Fig. 6.1 Illustrating the delay imposed by traditional runtime verification as well as existing proactive solutions with respect to our approach

400 types of operations [90], which implies 160,000 potential dependency relationship pairs may need to be examined by human experts.

- Furthermore, this only covers the static dependency relationships implied by the cloud design, whereas runtime patterns, e.g., those caused by business routines and user habits, cannot be captured in this way.
- Another critical limitation is that existing dependency models are deterministic in the sense that every event can only lead to a unique subsequent event. Therefore, the case demonstrated in the last two sessions $(N + 2, N + M)$ where the same event $(E3)$ may lead to several others $(E4$ or $E6)$ will not be captured by such models.

6.1.2   Threat Model

We assume that the cloud infrastructure management systems (1) may have implementation flaws, misconfigurations, and vulnerabilities that can be potentially exploited to violate security properties specified by the cloud tenants and (2) may be trusted for the integrity of the API calls, event notifications, logs, and database records (existing techniques on trusted computing may be applied to establish a chain of trust from TPM chips embedded inside the cloud hardware, e.g., [7, 58]). Though our framework may assist to avoid any violation of specified security properties due to either misconfigurations or exploits of vulnerabilities, our focus is not to detect specific attacks or intrusions. We focus on attacks directed through the cloud management interfaces (e.g., CLI, GUI), and any violation bypassing such interfaces is beyond the scope of this work. We assume a comprehensive list of critical events is provided upon which the accuracy of our auditing solution depends. However, we provide a guideline on identifying critical events in Sect. 6.7. Our proactive solution mainly targets certain security properties which would require a sequence of operations. To make our discussions more concrete, the following shows an example of in-scope threats based on a real-world vulnerability.

Running Example A real-world vulnerability in OpenStack,[1] CVE-2015-7713 [88], can be exploited to bypass security group rules (which are fine-grained, distributed security mechanisms in several cloud platforms including Amazon EC2, Microsoft Azure, and OpenStack to ensure isolation between instances). Figure 6.2 shows a potential deployment configuration, which might be exploited using this vulnerability. The pre-requisite steps of this scenario are to create VMA1 and VMB1 (*step 1*), create security groups A1 and B1 with two rules (i.e., *allow 1.10.0.7* and *allow 1.10.1.117*) (*step 2*), and start those VMs (*step 3*). Next, when Tenant A tries to delete one of the security rules (e.g., *allow 1.10.0.7*) (*step 4*), the rule is not removed from the security group of the active VMA1 due to the aforementioned

---

[1]OpenStack [90] is a popular open-source cloud infrastructure management platform.

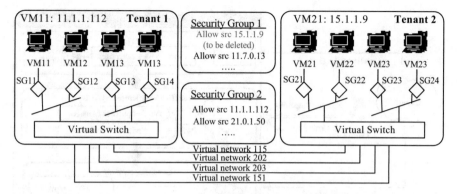

Fig. 6.2 An example scenario for exploiting a vulnerability in OpenStack [88], which leads to bypassing the security group mechanism

vulnerability. As a result, VMB1 is still able to reach VMA1 even though Tenant A intends to filter out that traffic. According to the vulnerability description, the security group bypass violation occurs only if this specific sequence of event instances (steps 1–4) happens in the mentioned order (namely, *event sequence*). In the next section, we present an overview of our approach and show how we automatically capture probabilistic dependencies among cloud events for proactive security auditing.

6.1.3 Approach Overview

In the following, we briefly describe our learning-based proactive auditing techniques used by LeaPS.

- First, it parses raw cloud logs into a structured format after marking each field of log entries so that log processing in the next step can be efficient.
- Next, it processes these parsed logs to interpret event types, aggregate log entries from different services (e.g., compute and network), and prepare inputs (as event sequences) for learning techniques.
- Then, it learns probabilistic dependencies between different event types captured as a Bayesian network from sequences of events processed from different cloud logs.
- Afterwards, LeaPS incrementally prepares the ground for runtime verification. The preparation is done based on the descending order of the conditional probability of critical events knowing any other given event has occurred.
- Finally, once one of these critical events is about to occur, we simply verify the parameters associated with its event instance with respect to the pre-computed conditions of that event and enforce the security policy according to the verification result.

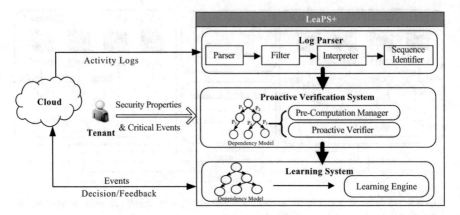

Fig. 6.3 An overview of LeaPS main components: log processing, learning, and auditing mechanisms

Figure 6.3 shows an overview of LeaPS. It consists of three major modules: log processor, learning system, and proactive verification system. The log processor is related to processing the unstructured and incomplete raw log files, which will be detailed in Sect. 6.2.1, and prepares the data to be used by the learning system. Our log processor consists of four major parts. The parser is responsible to identify fields for each log entries and parse them into a structured format. The filter extracts the relevant log entries and groups them based on tenant IDs. The interpreter is to mark event types for each log entry. Finally, the sequence identifier is responsible to extract the sequence out of those log entries and prepare inputs for various learning techniques. The learning system is dedicated for learning probabilistic dependencies for the model. The proactive verification system consists of two major parts. The pre-computation manager prepares the ground for the runtime verification. At runtime, a light-weight verification tool (e.g., proactive verifier [66]), which basically executes queries in the pre-computed results, is leveraged for the verification purpose. Based on the verification result, LeaPS provides a decision on the intercepted critical event instance.

6.2   Case Studies and Log Processing

In this section, we detail our approach for processing unstructured raw cloud logs and present the challenges and lessons learned from the analysis of the formats of real-world cloud logs in OpenStack [90].

6.2.1 Case Studies on Real-World Cloud Logs

As a first step, we conducted an investigation on how the executed operations at the cloud management level are logged in OpenStack [90], one of the major cloud management systems in today's cloud environments. To this end, we used two different OpenStack deployments: a real-life community cloud hosted at a real data center of a large telecommunication vendor and a cloud testbed managed within our institution. All sensitive information in the logs is anonymized based on the data owner's policies.

Background on OpenStack Logging Logging systems can provide an essential resource for both troubleshooting and behavior analysis of the underlying clouds. To this end, tracing back log entries to identify the root cause of a problem and subsequent actions, respectively, are natural solutions which motivate the processing of logs. Furthermore, the high complexity and ever-growing nature of cloud environments further increase the need for processing logs in a cloud. To this end, OpenStack [90] logs different user and system actions performed within the cloud. The most commonly used log format in OpenStack services starts with the timestamp, process ID, log level, the program generating the log, an ID, and followed by a message field, which might be divisible into smaller informative segments such as request method and URL Path. Furthermore, different OpenStack services write their log files to their corresponding subdirectory of the /var/log directory in the node they are installed in. For example, the log location of Nova is /var/log/nova. Figure 6.4 depicts an excerpt of the logs collected from a real cloud highlighting the useful information stored in these logs.

Investigated Factors In the following, we describe different factors in the logs that are relevant to automate the learning of dependency models in LeaPS.

1. **Layouts.** The first factor that we investigated is the general layout of logs. While comparing the layouts of the logs, we found that there are different attributes in each log entry, and those attributes vary based on the version and the logging service of OpenStack. This was exasperated by the

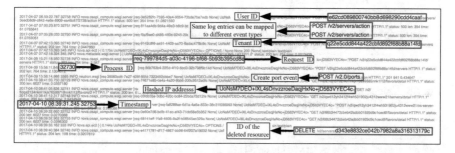

Fig. 6.4 Highlighting useful fields in one of the real-world cloud logs

lack of detailed documentation describing the meaning of these attributes. Through our study, we identified 11 fields that are used in OpenStack logging: timestamp, process-ID, log-type, method, request-ID, user-ID, tenant-ID, IP address, API URL, HTTP response, and request-length. Apart from common layout issues, we also observed discrepancies in layouts of the logs collected from the two different cloud deployments as each environment was managed by a different version of OpenStack.[2] The following example depicts the latter observation.

Figure 6.5 shows three examples of fields that are only present in one of the studied versions[3] of OpenStack: (1) The logs from the real cloud does not have any user IDs; instead, they store none; (2) the real cloud logs have entries starting with OPTIONS; and (3) the testbed log entries contain user-ID and tenant ID.

2. **Log Entries.** After identifying these differences in the layout, we scrutinize each log entry to enable understanding of the meaning of these entries and their related attributes. We observe that OpenStack logs a wide range of system-initiated events related to the coordination between different cloud services (e.g., compute, network, and storage). Such events are usually logged with a special tenant ID (tenant-service). The first row of Table 6.1 shows an example of such a log entry, where the ID of the tenant-service is dsfre23de8111214321def6e1e834re31. Moreover, there are requests to list resources or their corresponding details, which are made with GET in their method field. For instance, the second row of Table 6.1 shows a logged event for the tenant ID (77c433dsf43123edcc12349d9c16fcec) to render the Flavors (the component to show different resource consumptions by a VM). Both resource rendering and system-initiated logged events have no effect on changing cloud configurations, and therefore, these events are not useful for the auditing purpose.

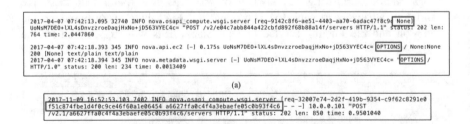

(a)

(b)

Fig. 6.5 Unique log fields and entries between two different studied versions of OpenStack: (a) real cloud and (b) testbed cloud

---

[2]We avoid disclosing the exact version details for the sake of security.

[3]Despite that the studied versions are directly consecutive, there are multiple differences in the logging system.

Table 6.1 Corresponding log entries to system initiated events in the real cloud

OpenStack log entry
"...POST /v2/dsfre23de8111214321def6e1e834re31/ os-server-external-events HTTP/1.1..."
"...GET /v2/77c433dsf43123edcc12349d9c16fcec/ flavors/detail HTTP/1.1..."

```
2017-04-10 08:41:03.750 32729 INFO nova.osapi_compute.wsgi.server [req-74671e85-ce4a-4a20-9bb6-200c9512a0fc
None] UoNsM7DEO+lXL4sDnvzzroeDaqjHxNo+jD563VYEC4c= "GET /v2 21djh3782nkml1kjk18299883nnk1l12 flavors/detail
HTTP/1.1" status: 200 len: 6027 time: 0.0357180
2017-04-10 08:41:05.626 32741 INFO nova.osapi_compute.wsgi.server [req-7036a44a-b539-4742-b01e-9b334905328f
None] UoNsM7DEO+lXL4sDnvzzroeDaqjHxNo+jD563VYEC4c= "GET /v2/
ffdqa2134nkml1kjk76598718nnk1l12 flavors/2 HTTP/1.1" status: 200 len: 615 time: 0.0270739
2017-04-10 08:41:12.317 32748 INFO nova.osapi_compute.wsgi.server [req-03fdb35a-3dbc-420c-bc84-22bda72d12fe
None] UoNsM7DEO+lXL4sDnvzzroeDaqjHxNo+jD563VYEC4c= "GET /v2 qwd12yh3412f4wh531902ju4312www21 servers/detail
HTTP/1.1" status: 200 len: 34955 time: 2.9434948
```

(a)

```
2018-02-06 12:00:15.013 4746 INFO nova.osapi_compute.wsgi.server [req-cf595608-f7dd-42a0-a476-e73fdcc8d89d
f51c874fbe1d4f0c9ce46f60a1e06454 a6627ffa0c4f4a3ebaefe05c0b93f4c6 - - -] 10.0.0.101 "DELETE /v2.1/
a6627ffa0c4f4a3ebaefe05c0b93f4c6/servers/73e39eec-53cb-4fa4-8938-9ecde73024de HTTP/1.1" status: 204 len: 274
time: 1.4679391
2018-02-06 12:01:34.190 4746 INFO nova.osapi_compute.wsgi.server [req-a3ec42b8-649c-4566-8549-620d8ec5576f
1125cc73a3c547d78c36eb12a98e5904 f72ea5b55e4c41f9917d9d8aa5c0f525 - - -] 10.0.0.101 "GET /v2.1/
f72ea5b55e4c41f9917d9d8aa5c0f525/flavors/3/os-extra_specs HTTP/1.1" status: 200 len: 286 time: 0.1976860
```

(b)

Fig. 6.6 Requests initiated by different tenants are logged into the same log file in both studied versions: (**a**) real cloud or (**b**) testbed cloud (the corresponding tenant IDs are highlighted)

Additionally, we notice that user-generated requests made under different tenants are jointly logged into the same log file. Furthermore, their log entries could be distinguished from each other based on the associated tenant ID field identifying the tenant initiating the request. Figure 6.6 highlights the different tenant IDs present in some entries within the log files of both real and testbed clouds.

3. **Type of Events.** In this part of the case studies, we investigate the process of identifying event types from user-generated requests. Usually, OpenStack user requests are transmitted to the server as REST API calls. Thus, our next step is to obtain the event type from each log entry. However, relying only on the REST methods (e.g., POST, GET, PUT, etc.) does not help as these methods do not map to a specific event type. Therefore, we study the API documentation of OpenStack to identify specific path information along with the REST methods (called URL path) to pinpoint each corresponding event type. Figure 6.7 shows examples of log entries highlighting URL paths corresponding to different event types. The URL paths in the figure actually refer to the event types *create VM*, *delete VM*, and *create port*, respectively.

However, there are event types for which we observe the same URL paths. For instance, Table 6.2 shows three examples of such URL paths from some log entries in the real cloud. Even though the three rows in the table correspond to three different events, their URL paths look identical except the VM ID field; which indicates that we cannot rely only on URL paths to uniquely

```
2017-04-10 08:39:22.021 32728 INFO nova.osapi_compute.wsgi.server [req-7de39040-457f-4e94-b137-d409d7c37173 None]
UoNsM7DE0+lXL4sDnvzzroeDaqjHxNo+jD563VYEC4c= "POST /v2/e04c7abb844a422cbfd892f68b88a14f/servers HTTP/1.1" status:
202 len: 764 time: 2.5971961

2017-04-10 08:54:55.384 32740 INFO nova.osapi_compute.wsgi.server [req-e4171781-df17-4867-bb96-64f207a19032 None]
UoNsM7DE0+lXL4sDnvzzroeDaqjHxNo+jD563VYEC4c= "DELETE /v2/e04c7abb844a422cbfd892f68b88a14f/servers/
d343e883-2ce0-42b7-982a-8a316313179c HTTP/1.1" status: 204 len: 198 time: 0.3207819
```
(a)

```
2017-10-21 18:13:34.319 7031 INFO neutron.wsgi [req-d16aeb53-c980-4ed7-9373-7df445afe585
96756ca976ba437fbe25671e41d0cd22 1d28697ef17540efba1677848a5b762a - - -] 10.0.0.102 - -
[21/Oct/2017 18:13:34] "POST /v2.0/ports.json HTTP/1.1" 201 1087 1.149421
```
(b)

Fig. 6.7 Logged REST API calls collected from two OpenStack versions deployed as (**a**) real cloud and (**b**) testbed cloud. The highlighted URL paths correspond to different cloud events (e.g., delete VM and create VM)

Table 6.2 Similar URL paths are logged for different cloud event types

OpenStack log entry	Event name
"POST /v2.1/a6627ffa0c4f4a3ebaefe05c0b93f4c6/servers/ f6128951-0c48-4a11-8b8b-5e96da77b698/"	Stop VM
"POST /v2.1/a6627ffa0c4f4a3ebaefe05c0b93f4c6/servers/ 1223d052-bc35-485a-9237-1830bca80fd7/"	Start VM
"POST /v2.1/a6627ffa0c4f4a3ebaefe05c0b93f4c6/servers/ 4c886192-43ad-4f98-90dd-34e24c84fcd0/"	Add security group

Table 6.3 Multiple entries are logged by different services following a user request

OpenStack log entry	Log file
2017-04-09 15:56:14.866 ... "POST /v2/e04c7abb844a422cbfd892f68b88a14f/servers HTTP/1.1"...	nova-api.log
2017-04-09 15:56:11.848 ... "POST /v2.0/ports.json HTTP/1.1"...	neutron-server.log

identify event type. As a result, using only those URL paths to identify their corresponding event types is insufficient.

4. **Correlations of Events.** Once event types of different log entries are identified, we need to investigate the relationships between these events and how they correlate. We find out that multiple log entries in different log files of different services correspond to the same user request, which implies that to complete certain user requests, OpenStack internally calls multiple APIs involving different services, thus generating multiple logged events. Table 6.3 shows an example, where the actual user request is to create a VM. However, we observe at least two entries (create VM and create port) in nova-api.log and neutron-server.log log files, respectively, to follow the initiation of the actual user request. Additionally, we notice that there are log entries in nova-api.log and neutron-server.log with the same timestamps (2017-11-01

18:17:16.345) corresponding to two different events (Add security group and Create port). Thus, distinguishing the right precedence relations between events cannot only rely on logged timestamps.

5. **Session Identification.** Finally, we need to split the log files into groups of events mainly based on the contexts (e.g., same user events) or session. The main intention behind this step is to prepare inputs for different learning techniques, many of which accept inputs as a sequence of events. However, in OpenStack logs, we observe that there exists no session-specific information. Moreover, most log entries do not include the requestor ID, which could have been useful to identify the context.

6.2.2   Real-World Challenges to Log Processing

In this section, we summarize the main challenges in processing cloud logs based on the above-mentioned study.

- **Heterogeneous Formats:** Our study shows that the cloud logs may have heterogeneous formats. The log formats vary within the same cloud platforms as well as for different versions of the management system, which includes varieties of different specifications (e.g., fields or attributes). In most cases, the logged information (fields) is not explicitly mentioned in the log file. As a result, it is non-trivial to interpret the log entries. Additionally, the log entries are not systematically generated. Therefore, processing such logs efficiently and systematically is challenging. To handle this challenge, we parse logs and store them in a structured manner by marking each attribute of logs. We will design a method for parsing in Sect. 6.2.3.2.
- **Ungrouped and Irrelevant Log Entries:** OpenStack logging system logs all tenants' requests jointly as well as system-initiated events, which are typically irrelevant to the auditing system. However, to detect/prevent security concerns at the tenant level, it is essential to separately analyze the log entries related to each tenant, and hence, current log formats are not appropriate for this purpose. Additionally, cloud generates many internal events to process a user request and stores log entries corresponding to those system-initiated events in the same log files. As a result, the analyses requiring to distinguish user actions from system actions can be hampered. This challenge will be tackled by grouping tenant-specific log entries and eliminating entries related to system-initiated events. More details are in Sect. 6.2.3.3.
- **Difficulties in the Identification of Event Types:** Identifying event types corresponding to each log entry is non-trivial due to the following reasons. First, while most clouds support REST APIs to request different operations (e.g., *create VM*, *create port*, and *update port*), the event type identification is not obvious from the API and requires checking the whole URL paths. Second, in some cases, URL paths for different events are the same, and hence, such paths are not

sufficient to identify these events uniquely. We call this kind of events *ambiguous* event. Ambiguous events will be properly identified by leveraging a special log option called `request body` in OpenStack. We detail the solutions to these challenges in Sect. 6.2.3.4.

- **Distributed Logging for Different Services:** Activities in different services (e.g., compute, network and storage) are logged separately. However, for log analysis, which involves different services combinedly such as ours, merging logs from these services is essential, but not trivial. There are certain requests that result in multiple log entries distributed over different log files by different services. Additionally, due to time synchronization issues between the services, multiple events may be logged with the same timestamp, which hinders analysis tools to extract the right precedence relationships between them. This challenge will be addressed by merging logs from different services and eliminating duplicate entries corresponding to the same user request. We provide more details of the solution in Sect. 6.2.3.5.

- **Need for a Sequence Identification Solution:** There exists no session-specific information in the logs. Also, the requester ID for a user request is missing in most log entries. Therefore, there is a need for a solution to identify sequences of events from these specific formats of logs fulfilling all the requirements (e.g., preserving transitions and their relative order) for a specific or a group of analyses purposes. To this end, we have already requested the OpenStack community to include requestor ID (at least) within each log entry to facilitate log analysis. Till then, we propose a custom algorithm to identify sequences of events, in which all transitions and their relative orders of the actual log entries are preserved. More details on this solution are presented in Sect. 6.2.3.6.

6.2.3 LeaPS Log Processing

In this section, we discuss our log processing approach, which addresses all of the above-mentioned challenges, and provides more structured and meaningful processed logs for different analyses. A high-level algorithm of our log processor is shown in Algorithm 2.

In the following, we briefly describe main steps of the log processing algorithm.

- **Line 3:** parses raw logs into a structured format. This step extracts identified fields in the log entries and uses them together with a set of pre-defined rules to parse the raw log into a structured log (e.g., CSV) file. This allows handling the heterogeneity of log formats.

- **Line 4:** groups parsed logs based on tenant IDs. The latter, easily identified in the obtained structured log file, allows grouping log entries based on the tenants under which the events are being logged. This tackles the issue of ungrouped events.

Algorithm 2 LeaPS log processing algorithm

Require: Parsing and matching rules for cloud events
Ensure: Event sequences for various log analyses (e.g., LeaPS and sequence pattern mining)
 1: **procedure** LOGPROCESSING(*CloudConfig*)
 2:    **for** each component ∈ *CloudConfig* **do**
 3:       Parse the raw logs;
 4:       Group parsed logs based on tenant IDs;
 5:       Prune irrelevant log entries (system-initiated and UI rendering);
 6:       Mark event types based on information in URL path and request body;
 7:    **end for**
 8:    Combine logs from different services (e.g., compute and network)
 9:    **for** each log entry ∈ *combinedLogs* **do**
10:       Identify *Sequences* from the combined logs
11:    **end for**
12:    return *Sequences*
13: **end procedure**

- **Line 5:** prunes irrelevant log entries. System-initiated log entries can be grouped and discarded easily based on the system tenant ID (i.e., tenant-service) present in each of the related log entries. Those related to the UI rendering actions are identified by inspecting the method used in the URL (e.g., GET) in the entries related to the logged API calls.
- **Line 6:** identifies the type of events for each log entry. We first identify event types from the method and path information available from Line 3. However, there are several event types (a.k.a. ambiguous events), which have the same method and path information. To tackle this, we further check the request body, which contains detailed information for each log entry, mainly by tuning logging options to include the missing information.
- **Line 7:** combines logs from different services (e.g., compute and network) based on different attributes (e.g., tenant id and request id) and timestamps. This step draws the correlation among events logged in different services so that it can handle the challenges mentioned in Sect. 6.2.2.
- **Lines 8–10:** construct the event sequences based on the occurrences of events in the actual log fulfilling the requirements mentioned in Sect. 6.2.2. Our log processor provides these event sequences as outputs, which can be later used by different analysis methods (e.g., LeaPS learning system in Sect. 6.3 and sequence pattern mining algorithms in Sect. 6.6.2).

Figure 6.8 illustrates an example of the outputs of each of these steps. In the following, we first describe the inputs to our log processing and then provide more details on each processing step.

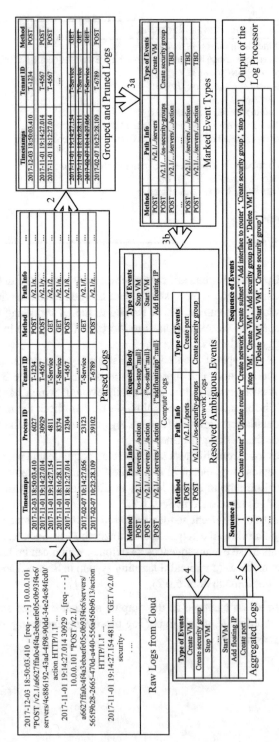

Fig. 6.8 Illustration of our log processor outputs after each step

6.2.3.1   Inputs to LeaPS Log Processing

Apart from raw cloud logs, our log processing algorithm requires two inputs: parsing rules to handle different log entries into a structured format and matching rules to identify the event types. Building these inputs is a one-time effort obtained from our investigations in the aforementioned case study.

Building Parsing Rules   To build the parsing rules, we first study different formats of logs and their corresponding structure. Next, we obtain different fields (e.g., timestamp, process ID and tenant ID) in the log entries and their relative positions in the logs. Finally, we build rules based on the fields and their corresponding orders in logs to support the parsing in Sect. 6.2.3.2.

Building Matching Rules   To identify the matching rules, we study the API documentation of OpenStack along with the log formats. To build a relationship between those fields in the logs and their corresponding event types. However, there exist several events for which all these fields are identical and hence, the event types of those events cannot be identified using this procedure. To tackle this, we leverage the request body, which contains detailed information for each log entry. Finally, we provide a complete mapping to identify different event types in Sect. 6.2.3.4.

6.2.3.2   Parsing Logs

The main purposes of this step are to mark all useful fields of different cloud logs and store them in a more structured way to enhance log analysis effort in terms of efficiency. We achieve these purposes, by parsing logs based on the pre-defined rules so that each identified field is marked with a meaningful name, and storing the logs in a more structured manner (e.g., in CSV) converting from a text-based file. For example, as shown in Step 1 of Fig. 6.8, we collect OpenStack logs from compute and network services. The first log entry   `2017-12-03 18:50:03.410 .. [req- - - -] 10.0.0.101 "POST /v2.1/a6627ffa0c4f4a3ebaefe05c0b93f4c6/servers/ 4c886192-43ad-4f98-90dd-34e24c84fcd0/ action HTTP/1.1"` contains the fields `timestamp`, `process ID`, `request ID`, `IP address`, `method`, `path info` and `tenant ID`, respectively. The next step is to parse each entry of the logs based on these fields and their relative positions, and to store them in a table (*Parsed Logs* in Fig. 6.8).

6.2.3.3   Grouping and Pruning of Parsed Logs

Extracting the contextual information, and separating both system- and user-initiated activities from the logs are the main goals of this step. First, we identify different contextual information such as tenant ID and accessed resource IDs (e.g., port ID, VM ID and subnet ID). Then, we group log entries based on the tenant

ID so that we can obtain tenant-specific activities together. Next, we extract the accessed resource IDs by the activities of all log entries. Finally, we separate log entries of user-initiated activities from that of system-initiated activities. Note that, the tenant ID field of the log entries for system-initiated events contains a special value (i.e., *tenant-service*) in OpenStack. However, there exist exceptions where system-initiated events are stored under an existing tenant. Therefore, we maintain a list of such exceptions and match them with the log entries while separating user-initiated events.

For instance, the *Parsed Logs* in Fig. 6.8 contain entries from tenants T-1234, T-4567, T-6789, and T-Service. After Step 2, in the *Grouped and Pruned Logs*, we first store all entries from the T-1234 tenant, and then, similarly store entries for T-4567 and T-6789. Afterward, we identify all log entries with the tenant ID T-Service and store them separately.

6.2.3.4 Marking Event Types

Marking the corresponding event types using the pre-defined matching rules (as shown) for each log entry is the main objective of this step. Based on the set of fields used in marking, there are two categories of event types. For the first category, we use the URL path (which includes method, resources, etc.). For example, from the *Grouped Logs* in Fig. 6.8, we identify method, resources, resource ID, and action as the potential fields, which together may provide unique information about each event type. Based on this assumption, we build the matching rules as shown in Table 6.4. The first entry in the table shows that the fields method: POST and resources: ports indicate the create port event

Table 6.4 Examples of mapping from URL-method and path-info to the event type

Method	Resources	Resource ID	Action	Event type
POST	Ports	NaN	NaN	Create port
PUT	Ports	Port ID	NaN	Update port
DELETE	Ports	Port ID	NaN	Delete port
PUT	Routers	Router ID	add_router _interface	Add interface to router
PUT	Routers	Router ID	remove_router _interface	Remove interface to router
POST	Floating-IPs	NaN	NaN	Create floating IP
PUT	Floating-IPs	VM ID	NaN	Associate floating IP
POST	Servers	VM ID	Actions	TBD

The term 'NaN' at fields of each row implies the absence of those fields in the corresponding log entry. The term 'TBD' implies that the corresponding event types are not identifiable according to the mentioned fields type. Whereas, to identify the add interface to router event type, we require the action: add_router_interface field along with the method, resources, and resource ID fields. However, for the last entry in the table, these fields are not sufficient to obtain the event type, as they have the same values for method, resources, resource ID, and action fields for multiple event types. These event types are considered as the second category and marked in the following manner.

To identify the second category of event types, we match the request ID in the log file and request body. For instance, Table 6.5 shows examples of matching rules between request body and event types. The first and second rows of the table show that the request body values {"os-start": null} and {"os-stop": null} ensure that the requested event types are start VM and stop VM, respectively. The third and fourth rows provide event types (Add security group and Add floating IP) along with their involved security group name (essential) and instance name (leapsVM), respectively.

In summary, we utilize the method, resource, resource ID, action, and request body fields to mark all event types.

6.2.3.5 Aggregating Logs

Merging logs from different services (e.g., compute, network, and storage) is the main goal of this step. To this end, we first combine multiple log files and sort them based on timestamp. Next, we identify entries with the same timestamp (if any) and mark them specially to later identify that they occurred at the same time in different services (if that helps any log analysis mechanism). Also, we identify any duplicate entry in different logs (as mentioned in Sect. 6.2.1 that the same event might be logged in multiple services) and only keep the corresponding log to the actual user request.

For example, Stage 5 in Table 6.6 shows two entries each from the compute service (Nova) and network service (Neutron), respectively. The first row of the table is for the Create VM event, which is actually initiated by a cloud user. On the

Table 6.5 Identifying event types from request bodies

Request body	Event type
req_body: {"os-start": null}	Start VM
req_body: {"os-stop": null}	Stop VM
req_body: {"addSecurityGroup": {"name": "essential"}}	Add security group
req_body: {"addFloatingIp": {"name": "leapsVM"}}	Add floating IP

Table 6.6 One user request (Create VM) is followed by multiple event logged by different OpenStack services, compute and network

OpenStack log entry	Event type	Initiated by

`"POST`
`/v2.1/a6627ffa0c4f4a3ebaefe05c0b93f4c6/`
`servers HTTP/1.1"`
	`Create VM`	`User`

`"POST /v2.0/ports.json HTTP/1.1"`	`Create Port`	`System`

other hand, the `Create port` event is a system-initiated event as a result of the user request. In other words, OpenStack creates a port by itself while creating a VM.

6.2.3.6 Generating Outputs

Our log processor provides outputs as sequences of events. In this work, we mainly observe the following three requirements for identifying events sequences. First, we preserve all transitions that are present in the actual logs. Second, we maintain the relative order between events. Third, in each sequence, we avoid cycles (by starting a new sequence when there is a repetition) to facilitate capturing relationships between events (e.g., dependencies in our model), flowing from top to bottom. To validate our approach, we use the generated events sequences to build a Bayesian network and to perform sequence pattern mining in Sects. 6.3 and 6.6.2, respectively.

To generate the final output (i.e., sequences of events), the log processor performs the following steps:

- Read event types sorted by timestamp in the *Aggregated Logs*, and group event types in a sequence till any event type is observed for the second time. In other words, a sequence Seq_i contains all event types from $Event_m$ to $Event_{n-1}$, where the $Event_n$ (the successor of $Event_{n-1}$) has already been observed in the sequence Seq_i. Thus, no sequence contains any repeated event types, and hence, we avoid cycles in sequences.
- Start the next sequence from the last element of the previous sequence so that all transitions within the sequences are preserved. In other words, the following sequence, Seq_{i+1}, starts with the $Event_{n-1}$, which is the last event of the Seq_i sequence.

In summary, our log processing approach addresses all challenges discussed in Sect. 6.2.2. As an example, it provides sequences of events as shown in Table 6.7. Later, these sequences will be utilized by our learning system to learn the dependency model presented in the next section. The implementation details including the algorithms for our log processing are presented in Sect. 6.5.3. Also, the performance evaluation of our log processor is shown in Sect. 6.6.2.

Table 6.7 Some examples of LeaPS log processor outputs

Aggregated log content
{Create router, Update router, Create network, Create VM, Add interface to router, Create security group, Start VM, Create VM, Add security group rule, Delete VM, Create VM, Create security group, Start VM}

Output
Seq_1 = {Create router, Update router, Create network, Create VM, Add interface to router, Create security group, Start VM}
Seq_2 = {Start VM, Create VM, Add security group rule, Delete VM}
Seq_3 = {Delete VM, Create VM, Create security group, Start VM}

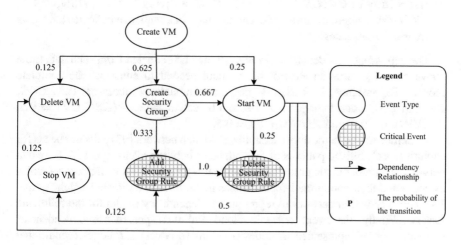

Fig. 6.9 An excerpt of a dependency model in the form of a Bayesian network

6.3 LeaPS Learning System

This section first describes the dependency model and then presents the steps to learn probabilistic dependencies for this model.

6.3.1 The Dependency Model

We first demonstrate our dependency model through an example and then formally define the model. The model will be the foundation of our proactive auditing solution (detailed in Sect. 6.4).

Figure 6.9 shows an example of a dependency model, where nodes represent different event types in a cloud and edges represent transitions between event types. For example, nodes, *create VM*, and *create security group* represent the corresponding event types, and the edge from *create VM* to *create security group* indicates the likely order of occurrence of those event types. The label of this edge, 0.625, means 62.5% of the times an instance of the *create VM* event type will be followed by an instance of the *create security group* event type.

Our objective is to automatically construct such a model from logs in clouds. As an example, the following shows an excerpt of the event types *event-type* and historical event sequences *hist* for 4 days related to the running example of Sect. 6.1.2.

- *event-type* = {*create VM (CV), create security group (CSG), start VM (SV), delete security group rule (DSG)*}; and
- $hist = \{day\ 1 : CV, CSG, SV;\ day\ 2 : CSG, SV;\ day\ 3 : CSG, DSG;\ day\ 4 : CV, DSG\}$, where the order of event instances in a sequence indicates the actual order of occurrences.

The dependency model shown in Fig. 6.9 may be extracted from such data (note above we only show an excerpt of the data needed to construct the complete model). For instance, in $hist$, CV has three immediate successors (i.e., CSG, SV, and DSG), and their probabilities can be calculated as $P(CSG|CV) = 0.5$, $P(SV|CV) = 0.5$, and $P(DSG|CV) = 0.5$.

As demonstrated in the above example, Bayesian network [93] suits our needs for capturing probabilistic patterns of dependencies between events types. A Bayesian network is a probabilistic graphical model that represents a set of random variables as nodes and their conditional dependencies in the form of a directed acyclic graph. We choose Bayesian network to represent our dependency model for the following reasons. Firstly, the event types in cloud and their precedence dependencies can naturally be represented as nodes (random variables) and edges (conditional dependencies) of a Bayesian network. Secondly, the need of our approach for learning the conditional dependencies can be easily implemented as parameter learning in Bayesian network. For instance, in Fig. 6.9, using the Bayes' theorem we can calculate the probability for an instance of *add security group rule* to occur after observing an instance of *create VM* to be 0.52. More formally, the following defines our dependency model.

Given a list of event types *event-type* and the log of historical events *hist*, the *dependency model* is defined as a Bayesian network $B = (G, \theta)$, where G is a DAG in which each node corresponds to an event type in *event-type*, and each directed edge between two nodes indicates the first node would immediately precede the other in some event sequences in *hist* whose probability is part of the list of parameters, θ.

We say a *dependency* exists between any two event types if their corresponding nodes are connected by an edge in the dependency model, and we say they are not dependent, otherwise. We assume a subset of the leaf nodes in the dependency model is given as *critical events* that might breach some given *security properties*.

6.3.2 Learning Engine

The next step is to learn the probabilistic dependency model from the sequences of event instances in the processed logs. To this end, we choose the parameter learning technique in Bayesian network [45, 74, 93] (this choice has been justified in Sect. 6.3.1). We now first demonstrate the learning steps through an example and then provide further details.

Figure 6.10 shows the dependency model of Fig. 6.9 with the outcomes of different learning steps as the labels of edges. The first learning step is to define a priori, where the nodes represent the set of event types received as input, and the edges represent possible transitions from an event type, e.g., from the *create VM* event to the *delete VM, start VM*, and *create security group* events. Then, $P(DV|CV)$, $P(CSG|CV)$, $P(SV|CV)$, and other conditional probabilities (between immediately adjacent nodes in the model) are the parameters; all parameters are initialized with equal probabilities. For instance, we use 0.33 to label each of the three outgoing edges from the *create VM* node. The second learning step is to use the historical data to train the model. For instance, the second values in the labels of the edges of Fig. 6.10 are learned from the processed logs obtained from the log processor. The third values in the labels of Fig. 6.10 represent an incremental update of the learned model using the feedback from a sequence of runtime events.

This learning mechanism mainly takes two inputs: the structure of the model with its parameters and the historical data. The structure of the model, meaning the nodes and edges in a Bayesian network, is first derived from the set of event types received as input. To this end, we provide a guideline on identifying such a set of event types in Sect. 6.7. Initially, the system considers every possible edge between nodes (and eventually deletes the edges with probability 0), and conditional probabilities between immediately adjacent nodes (measured as the conditional probability) are chosen as the parameters of the model. We further sparse the structure into smaller groups based on different security properties (the structure in Fig. 6.10 is one of the

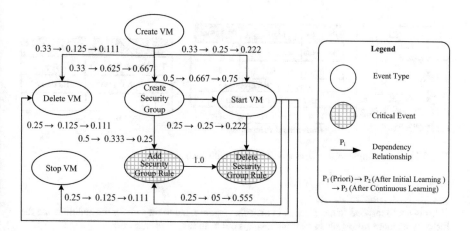

Fig. 6.10 Parameters of the dependency model are learned through multiple steps

examples). The processed logs containing sequences of event instances serve as the input data to the learning engine for learning the parameters. Finally, the parameter learning in Bayesian network is performed as follows: (1) defining a priori (with the structure and initialized parameters of the model), (2) training the initial model based on the historical data, and (3) continuously updating the learned model based on incremental feedbacks.

6.4 LeaPS Proactive Verification System

This section presents our learning-based proactive verification system.

6.4.1 Likelihood Evaluator

The likelihood evaluator is mainly responsible for triggering the pre-computation. To this end, the evaluator first takes the learned dependency model as input, and derives offline all indirect dependency relationships for each node. Based on these dependency relationships, the evaluator identifies the event types for which an immediate pre-computation is required. Additionally, at runtime the evaluator matches the intercepted event instance with the event type and decides whether to trigger a pre-computation or verification request.[4] The data manipulated by the likelihood evaluator based on the dependency model will be described using the following example.

Figure 6.11 shows an excerpt of the steps and their outputs in the likelihood evaluator module. In this figure, the *Property-CE-Threshold* table maps the *no bypass of security group* property [17] with its critical events (i.e., *add security group rule* and *delete security group rule*) and corresponding thresholds (i.e., 0.5

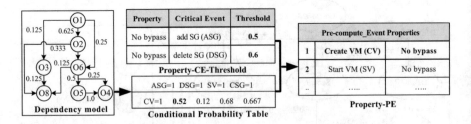

Fig. 6.11 An example of the likelihood evaluator with each step outputs

---

[4]This is not to respond to the event as in incident response, but to prepare for the auditing, and the incident response following an auditing result is out of the scope of this work.

and 0.6). Then, from the conditional probability in the model, the evaluator infers conditional probabilities of all possible successors (both direct and indirect) and stores them in the *Conditional-Probability* table. The conditional probability for *ASG* having *CV* (i.e., $P(ASG|CV)$) is 0.52 in the *Conditional-Probability* table in Fig. 6.11. Next, this value is compared with the thresholds of the *no bypass* property in the *Property-CE-threshold* table. As the reported probability is higher, the *CV* event type is stored in the *Property-PE* table so that for the next *CV* event instance, the evaluator triggers a pre-computation.

6.4.2 Pre-computing Module

The purpose of the pre-computing module is to prepare the ground for the runtime verification. In this work, we mainly discuss watchlist-based pre-computation [66]; where watchlist is a list containing all allowed parameters for different critical events. The specification of contents in a watchlist is defined by the cloud tenant and is stored in the *Property-WL* table. We assume that at the time LeaPS is launched, we initialize several tables based on the cloud context and tenant inputs. For instance, inputs including the list of security properties, their corresponding critical events, and the specification of contents in watchlists are first stored in the *Property-WL* and *Property-CE-Threshold* tables. The watchlists are also populated from the current cloud context. We maintain a watchlist for each security property. Afterwards, each time the pre-computation is triggered by the likelihood evaluator, this module incrementally updates the watchlist based on the changes applied to the cloud in the meantime. The main functionality of the pre-computing module is described using the following example.

Left side of Fig. 6.12 shows two inputs (*Property-WL* and *Property-PE* tables) to the pre-computing module. We now simulate a sequence of intercepted events (shown in the middle box of the figure) and depict the evolution of a watchlist for the *no bypass* property (right side box of the figure). (1) We intercept the *create VM 1733* event instance, identify the event in the *Property-PE* table, and add VM 1733

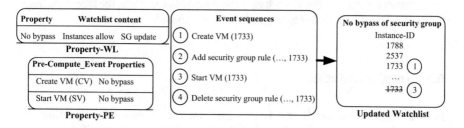

Fig. 6.12 Steps of updating the watchlist content through a sample sequence of events

to the watchlist without blocking it. (2) After intercepting the *add security group rule (..., 1733)* event instance, we identify that this is a critical event. Therefore, we verify with the watchlist keeping the operation blocked. We find that VM 1733 is in the watchlist, and hence, we recommend to allow this operation. (3) We intercept the *start VM 1733* operation and identify the event in the *Property-PE* table. VM 1733 is then removed from the watchlist, as the VM is active. (4) After intercepting the *delete security group rule (..., 1733)* event instance, we identify that this is a critical event. Therefore, we verify with the watchlist keeping the event instance blocked, find that VM 1733 is not in the watchlist, and hence, identify the current situation as a violation of the *no bypass* property.

6.4.3   Feedback Module

The main purposes of the feedback module are: (1) to provide feedback to the learning engine and (2) to provide feedback to the tenant on thresholds for different properties. These purposes are achieved by three steps: storing verification results in the repository, analyzing the results, and providing the necessary feedback to corresponding modules.

Firstly, the feedback module stores the verification results in the repository. Additionally, this module stores the verification result as hit or miss after each critical event, where the hit means the requested parameter is present in the watchlist (meaning no violation) and the miss means the requested parameter is not found in the watchlist (meaning a violation). Additionally, we store the sequence of events for a particular time period (e.g., one day) in a similar format as the processed log described in the learning module. In the next step, we analyze these results along with the models to prepare a feedback for different modules. From the sequence of events, the analyzer identifies whether the pattern is already observed or is a new trend, and accordingly the updater prepares a feedback for the learning engine either to fine-tune the parameter or to capture a new trend. From the verification results, the analyzer counts the number of misses for different properties to provide a feedback to the user on their choice of thresholds (stored in the *Property-CE-Threshold* table) for different properties. For more frequently violated properties, the threshold might be set to a lower probability to trigger the pre-computation earlier.

6.5   Implementation

In this section, we detail the LeaPS implementation and its integration into OpenStack along with the architecture of LeaPS (Fig. 6.13) and a detailed algorithm.

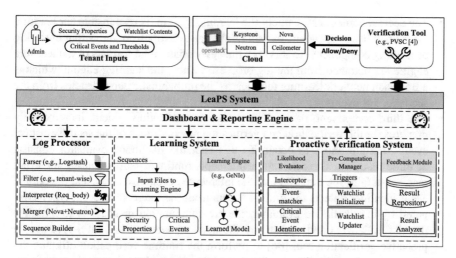

Fig. 6.13 LeaPS auditing system architecture

6.5.1 Background

OpenStack [90] is an open-source cloud management platform that is being used almost in half of private clouds and significant portions of the public clouds (see [92] for detailed statistics). Neutron is its network component, Nova is its compute component, and Ceilometer is its telemetry for receiving event histories from other components. Each component of OpenStack generates notifications, which are triggered by pre-defined activities such as VM creation, security group addition, and are sent to Ceilometer for monitoring purposes. Ceilometer extracts the information from the notifications and transforms them into events.

6.5.2 LeaPS Architecture

Figure 6.13 shows an architecture of LeaPS. It has four main components: log processor, learning system, verification system, and dashboard & reporting engine.

- The first component, namely the log processor, obtains sequences of events from the retrieved raw cloud logs. To this end, the parser module first processes the raw logs to retrieve identified fields in each log entry and systematically generates structured content stored as CSV files. Then, the filter module groups the log entries tenant-wise and separates these log entries corresponding to the user-initiated events. The interpreter module consults the mapping of URL paths and request body to identify the corresponding event type of a log entry. The merger module combines logs from different services (e.g., Neutron and Nova) of Open-Stack. The sequence builder generates the sequences of events from the logs.

- The second component, namely the learning system, is responsible for learning the probabilistic dependencies using Bayesian network from the output of the log processing component. To this end, the appropriate input formats of the learning engine are obtained from the log processor. Then, the learning engine, which is a Bayesian network learning tool, learns the probabilistic dependencies from the sequences of events.
- The third component, namely the proactive verification system, incrementally prepares for the verification and verifies the preconditions of the security critical events that are about to occur. To this end, the likelihood evaluator consists of three modules. The interceptor intercepts runtime event instances, the event matcher obtains the event type of the intercepted event instances, and the critical event identifier detects the critical events from the intercepted event type. Triggered by the likelihood evaluator, the pre-computation manager is to initialize (by the initializer) and update (by the updater) watchlists. LeaPS leverages a proactive verification tool [66] to perform the runtime verification utilizing the pre-computed results. The feedback module is to analyze the previous verification results and to provide feedback to update the probabilities in the model.
- The fourth component, namely the dashboard and reporting engine, is to provide an interface to LeaPS users to interact with the system and to observe different verification results.

In the following, we describe the implementation details of different components of LeaPS.

6.5.3   Log Processor

The log processor first automatically collects logs from different OpenStack components, e.g., Nova, Neutron, Ceilometer, etc. We use Logstash [30], a popular open-source data processing tool, for transforming unstructured and semi-structured logs into CSV format and available for further processing. To enable Logstash transformation, we use the parsing rules that we build for OpenStack logs in our case study. Afterwards, we implement filters in Python to group and eliminate log entries. Then, we build a mapping between URL paths with request body and event types and consult this map to identify the event type of each log entry. Next, we merge Neutron and Nova logs based on the timestamps while handling conflicting issues. For example, while a user requests to create a VM, the event (i.e., create port) happening at Neutron is done by the `tenant-service` and is removed while dividing events into different tenant groups. Finally, to prepare the logs to be used in the LeaPS learning system and for other log analysis purposes, we run a custom algorithm, which preserves all transitions in the actual logs, implemented in Python to identify sequences in combined logs. The *processLogs* procedure in Algorithm 3 implements all above-mentioned steps of our log processor.

Algorithm 3 Log processing ()

Require: *CloudOS, parsing-rules, matching-rules*
Ensure: *sequence[]*
 1: **procedure** PROCESSLOGS(*CloudOS*)
 2:   **for** each component $c_i \in CloudOS$ **do**
 3:     $rawLogs = \text{collectLog}(c_i)$
 4:     $Fields[] = \text{identifyLogFields}(rawLogs, parsing\text{-}rules)$
 5:     $parsedLogs[c_i] = \text{parseLogs}(rawLogs, Fields[], parsing\text{-}rules)$
 6:     $systemEvents[] = \text{identifySystemEvents}(CloudOS)$
 7:     $prunedLogs[c_i] = \text{pruneLogs}(parsedLogs[c_i], systemEvents[])$
 8:     $groupedLogs[c_i] = \text{groupLogs}(prunedLogs[c_i], CloudOS.tenants[])$
 9:     $markedEvents[c_i] = \text{markEvents}(groupedLogs[c_i], matching\text{-}rules)$
10:   **end for**
11:   $combinedLogs = \text{combineLogs}(markedEvents[])$
12:   **for** each log entry $entry_i \in combinedLogs$ **do**
13:     **if** $entry_i$ is not already in $sequence[j]$ **then**
14:       $sequence[j] = sequence[j] + entry_i$
15:     **else**
16:       $j{+}{+};$
17:       $sequence[j] = sequence[j] + entry_i$
18:     **end if**
19:   **end for**
20:   return $sequence[]$
21: **end procedure**

6.5.4 Learning System

For learning, we leverage SMILE and GeNIe [6], which is a popular tool for modeling and learning with Bayesian network. SMILE and GeNIe uses the EM algorithm [24, 57] for parameter learning. The learning module is responsible for preparing inputs to GeNIe and conducting the learning process using GeNIe. The sequences obtained from the log processor are further processed to convert them into the input format (in .dat) of GeNIe. Additionally, the structure of the Bayesian network and its parameters are provided to GeNIe. Furthermore, we choose the uniform option, where the assumption is that all parameters in the network form the uniform distribution with confidence equal to one. Finally, GeNIe provides an estimation of the parameters, which are basically probabilities of different transitions in the dependency model. Additionally, to execute sequence pattern mining algorithms with log processor outputs, we leverage SPMF [35], which is a popular open-source data mining library. The *Learn* procedure in Algorithm 4 implements the learning steps of LeaPS.

Algorithm 4 Learning-based proactive verification ()

Require: *CloudOS, Properties, structure, sequence*[]
Ensure: *decision*
 1: **procedure** LEARN(*sequence*[], *structure, Properties*)
 2:     **for** each property $p_i \in Properties$ **do**
 3:         *learnedParameters* = learnModel(*structure, sequence*[])
 4:         *dependencyModel* = buildModel(*structure, learnedParameters, p_i.critical-events*)
 5:     **end for**return *dependencyModels*
 6: **end procedure**

 7: **procedure** EVALUATE-LIKELIHOOD(*CloudOS, Properties, dependencyModels*)
 8:     **for** each event type $e_i \in CloudOS.event$ **do**
 9:         *Conditional-Probability-Table* = inferLikelihood(e_i, *dependencyModels*)
10:         **if**   checkThreshold(*Conditional-Probability-Table,Property-CE-Threshold*) **then**
11:             insertProperty PE(e_i,*Property CE Threshold.property*)
12:         **end if**
13:     **end for**
14:     *interceptedEvent* = intercept-and-match(*CloudOS, Event-Operation*)
15:     **if** *interceptedEvent* $\in$ *Properties.critical-events* **then**
16:         *decision* = verifyWL(*Properties.WL, interceptedEvent.params*)
17:         return *decision*
18:     **else if** *interceptedEvent* $\in$ *Property-PE* **then**
19:         Pre-compute-update(*Properties, interceptedEvent.params*)
20:     **end if**
21: **end procedure**

22: **procedure** PRE-COMPUTE-INITIALIZE(*CloudOS, Properties*)
23:     **for** each property $p_i \in Properties$ **do**
24:         WL_i= initializeWatchlist($p_i.WL, CloudOS$)
25:     **end for**
26: **end procedure**
27: **procedure** PRE-COMPUTE-UPDATE(*Properties, parameters*)
28:     updateWatchlist(*Properties.WL, parameters*)
29: **end procedure**

30: **procedure** FEEDBACK(*Result, dependencyModels, Properties*)
31:     storeResults(*Result, dependencyModels*)
32:     **if** analyzeSequence(*Result.seq*) = "new-trend" **then**
33:         updateModel(*Result.seq*,'new')
34:     **else**
35:         updateModel(*Result.seq*,'old')
36:     **end if**
37:     **for** each property $p_i \in Properties$ **do**
38:         *change-in-threshold*[i] = analyzeDecision(*Result.decision, p_i*)
39:     **end for**
40: **end procedure**

6.5.5 Proactive Verification System

We intercept all requests to the Nova service as they are passed through the Nova pipeline, having the LeaPS middleware inserted in the pipeline. The body of requests, contained in the wsgi.input attribute of the intercepted requests, is scrutinized to identify the type of requested events. Next, the pre-computing module stores the result of inspection in a MySQL database. The feedback module is implemented in Python. Those modules work together to support the methodology described in Sect. 6.4, as detailed in Algorithm 4.

6.5.6 Dashboard and Reporting Engine

LeaPS users interact with the system through a dashboard, which is implemented using a web interface in PHP. Through this dashboard, users can enable proactive auditing so that LeaPS starts intercepting cloud events and verifies them. In the dashboard, tenant admins can initially select security properties from different standards. (e.g., ISO 27017, CCM V3.0.1, NIST 800-53, etc.) Through the monitoring panel, LeaPS continuously updates the summary of the verification results. Furthermore, the details of any violation with a list of evidence are also provided. Moreover, our reporting engine archives all the verification reports for a pre-defined period.

6.6 Experimental Results

In this section, we first describe the experiment settings and then present LeaPS experimental results with both synthetic and real data.

6.6.1 Experimental Setting

Both experiments on LeaPS log processor and proactive verification system involve datasets collected from our testbed and the real cloud. In the following, we describe both environmental settings.

Testbed Cloud Settings Our testbed cloud is based on OpenStack version Mitaka. There are one controller node and up to 80 compute nodes, each having Intel i7 dual core CPU and 2 GB memory running Ubuntu 16.04 server. Based on a recent survey [92] on OpenStack, we simulate an environment with maximum 1000 tenants and 100,000 VMs. We conduct the experiment for 10 different datasets varying the number of tenants from 100 to 1000 while keeping the number of VMs fixed to 1000 per tenant. For Bayesian network learning, we use GeNIe academic version 2.1. For

Table 6.8 The number of
events generated in our
testbed cloud and logged by
Neutron and Nova services
for different datasets

Dataset	Nova	Neutron
DS1	9997	7998
DS2	20,000	15,998
DS3	29,998	23,999
DS4	39,998	32,000
DS5	48,995	40,293

sequential pattern mining, we use SPMF v.2.20. Table 6.8 describes the datasets for
experiments on log processing. We repeat each experiment 100 times.

Real Cloud Settings We further test LeaPS using data collected from a real
community cloud hosted at one of the largest telecommunication vendors. To this
end, we analyze the management logs (sized more than 1.6 GB text-based logs)
and extract 128,264 relevant log entries for the period of more than 500 days. As
Ceilometer is not configured in this cloud, we utilize Nova and Neutron logs, which
increases the log processing efforts significantly.

6.6.2   Results on Log Processor

In the following, we present obtained experiment results for our log processor both
in testbed and real clouds.

Experiments with Testbed Cloud   The objective of the first set of the experiments
is to measure the efficiency of our log processor for two different cloud services,
e.g., compute (Nova) and network (Neutron). Figure 6.14 shows the time required
(in seconds) to parse logs of different datasets. The results show that the parsing is
the most time consuming step in log processing, as this step parses text-based logs
and stores them into CSV files. The parsing of our largest dataset (DS5) requires
around 3 min and around 2 min for Nova and Neutron logs, respectively. Figure 6.15
presents the results (in seconds) to interpret event types of all entries from the
grouped logs for Nova and Neutron. The trend for both services shows almost a
linear increase while varying the number of log entries. Interpreting event types
for the largest dataset takes 16.72 and 11.14 s for Nova and Neutron, respectively.
Figure 6.16 shows the time required (in seconds) to group the log entries based on
tenant IDs and to eliminate system-initiated entries (from tenant-service). For the
largest dataset of Nova, the required time remains within 80 ms. Grouping Neutron
logs, which is comparatively smaller in size, requires maximum 55 ms.

The second set of experiments is to measure the efficiency of aggregating logs
from different services and generating outputs by our log processor. Figure 6.17
shows the required time (in seconds) to aggregate logs from Nova and Neutron
and to generate inputs to the sequence pattern mining algorithms implemented in

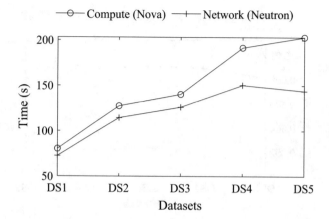

Fig. 6.14 The required time (in seconds) for parsing raw logs with respect to the number of provided events in different datasets

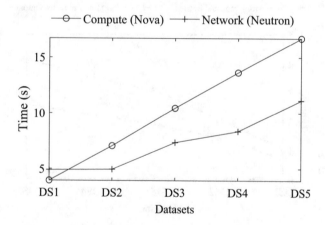

Fig. 6.15 The required time (in seconds) for interpreting event types with respect to the number of provided events in different datasets

Fig. 6.16 The required time (in Seconds) to group log entries for each tenant while varying the number of events provided in different datasets

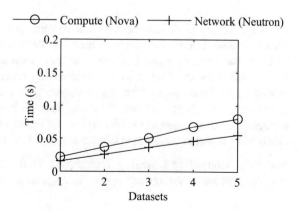

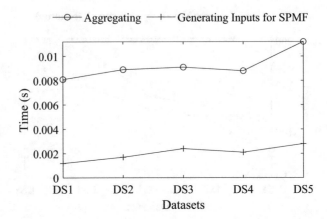

Fig. 6.17 The required time (in seconds) to merge logs of nova_api and neutron_server and to provide inputs for the pattern mining library (SPMF)

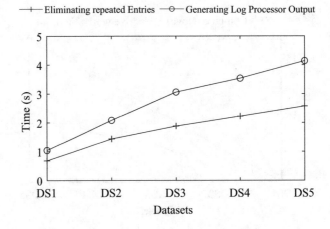

Fig. 6.18 The required time (in seconds) for eliminating the immediately repeated entries with respect to the number of provided events in different datasets

the SPMF library [35]. The required time to aggregate logs remains within 112 ms for the largest dataset. The required time for the input generation remains within 2 ms for the largest dataset and shows a linear increase. Figure 6.18 depicts the time required for the final steps, i.e., eliminating repeated entries and identifying sequences as the outputs of the log processor, performed on the aggregated logs. In both steps, the larger datasets show less increase in the required time than other datasets. The time required for identifying sequences remains within 2.6 s, and the eliminating repeated entries step takes maximum 4.2 s for our largest dataset.

Applying Alternative Learning Techniques To demonstrate the applicability of our log processor, we further apply its outputs to run three popular sequence pattern mining (which is a data mining approach with broad range of applications) algorithms. Specifically, we first run the PrefixSpan [94] algorithm, which mines the complete set of patterns. One potential application could be to identify structures of dependencies among cloud events using this algorithm to further enhance the proactive security solutions. Second, we execute the MaxSP [36] algorithm, which mines maximal sequential patterns. Using this algorithm, we can easily identify the most common patterns, which potentially can facilitate anomaly-based security solutions. Third, we run the ClaSP [39] algorithm, which identifies the largest pattern with a minimum frequency. This algorithm might be useful to identify unique patterns to profile a security violation or a legitimate use. To this end, we generate inputs to SPMF, which is a sequence pattern mining tool, and report the input generation time in Fig. 6.18. We also report the efficiency results to run these algorithms with the outputs of our log processor in Fig. 6.19. We observe constant time (1 ms) while running ClaSP algorithm for different datasets. The MaxSP and PrefixSpan algorithms take around 18 and 6.5 ms, respectively, for our largest dataset.

Experiments with Real Cloud The main objective of this part of the experiments is to evaluate the applicability of our log processor in a real cloud environment. Table 6.9 shows the summary of the results that we obtain for the real data. Due to the much larger size (e.g., 1.6 GB text-based logs) of the real-life logs, the parsing time is quite long (4 h and 40 min). However, once LeaPS is active, it may potentially log intercepted events in an incremental manner to avoid the delays at the parsing step. After parsing, we eliminate the log entries related to listing resources and their details, as the corresponding events to these entries are beyond our interest. The time for the remaining steps is quite similar to what is measured for our testbed cloud logs with much smaller size of logs. Note that the grouping step is not measured for Neutron, as the tenant ID is missing in the Neutron logs collected from the real cloud (as discussed in Sect. 6.2.1). However, we group them arbitrarily to measure the time for next steps. From the results of the real data, our observation is that our

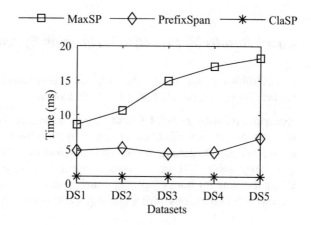

Fig. 6.19 The required time (in seconds) for running different pattern mining algorithms in the SPMF library: PrefixSpan, MaxSP, and ClaSP

Table 6.9 The required time to perform different steps of experiments on real data: Nova and Neutron logged events

Services	# of log entries	Parsing	Grouping	Interpretation	Merging	Generating sequences
Nova	1,450,011	4 h 40 m	0.0777 s	99.271 s	0.02206 s	1.4483 s
Neutron	3,992,644		–	51.820 s		

The time duration of parsing, merging, and generating sequences steps are measured for the aggregated logged entries of these services. As the tenant ID is not included in the Neutron logs collected from the real cloud, the grouping step is not measured for these entries

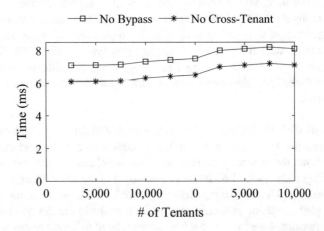

Fig. 6.20 The required time for the online runtime verification with respect to different number of VMs for the *no bypass* and *no cross-tenant* properties. The imposed delay consists of the time to perform interception, matching of event type, and checking in the watchlist

log processor is scalable once the parsing step is performed, which possibly allows our approach to process huge logs in a reasonable time.

6.6.3   Results on Proactive Verification System

In the following, we discuss the obtained experimental results for our proactive verification system both in testbed and real clouds.

Experiments with Testbed Cloud The objective of the first set of experiments with our proactive verification system is to demonstrate the time efficiency. Figure 6.20 shows the time in milliseconds required by LeaPS to verify the *no bypass of security group* [17] and *no cross-tenant port* [52] properties. Our experiment shows the time for both properties remains almost the same for different datasets because most operations during this step are database queries; SQL queries for our different datasets almost take the same time. Figure 6.21 shows the time (in seconds) required by GeNIe to learn the model while we vary the number of events from 2000 to 10,000. In Fig. 6.22, we measure the time required for different steps of the offline pre-computation for the *no bypass* property. The total time (including the time of incrementally updating WL and updating PE) required for the largest dataset is about 8 s, which justifies performing the pre-computation proactively. The one-time initialization of pre-computation is performed in 50 s for the largest dataset. Figure 6.23 shows the time in seconds required to update the model and to update the list of pre-compute events. In total, LeaPS requires less than 3.5 s for this step.

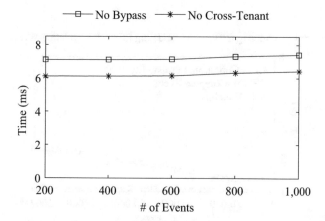

Fig. 6.21 The required time for the offline learning process with respect to different number of event instances in the logs for the *no bypass* and *no cross-tenant* properties

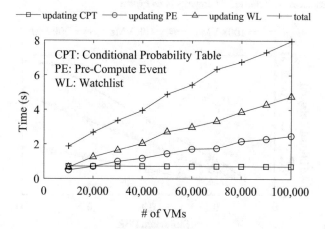

Fig. 6.22 The required pre-computation time with respect to different number of instances. The considered security property is *no bypass*

In the second set of experiments, we demonstrate how much LeaPS may be affected by a wrong prediction resulted from inaccurate learning. For this experiment, we simulate different prediction error rates (PER) of a learning engine ranging from 0 to 0.4 on the likelihood evaluator procedure in Algorithm 4. Figure 6.24 shows in seconds the additional delay in the pre-computation caused by the different PER of a learning engine for three different number of VMs. Note that, the pre-computation in LeaPS is an offline step. The delay caused by 40% PER for up to 100k VMs remains under 2 s, which is still acceptable for most applications.

In the final set of experiments, we compare LeaPS with a baseline approach (similar to [66]), where all possible paths are considered with equal weight, and the number of steps in the model is the deciding factor for triggering the pre-

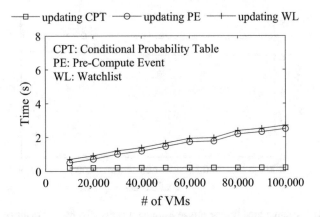

Fig. 6.23 The required time (in seconds) for the feedback modules with respect to different number of instances. The considered property is *no bypass*

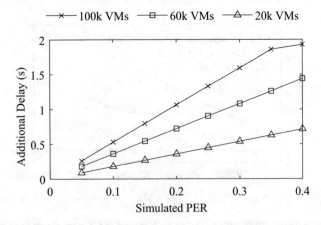

Fig. 6.24 The effect (in seconds) of (simulated) prediction error rates (PER) of a learning tool on pre-computation time

computation. Figure 6.25 shows the pre-computation time for both approaches in the average case, and LeaPS performs about 50% faster than the baseline approach (the main reason is that, in contrast to the baseline, LeaPS avoids the pre-computation for half of the critical events on average by leveraging the probabilistic dependency model). For this experiment, we choose the threshold, N-th (an input to the baseline), as two and the number of security properties as four. Increasing both the value of N-th and the number of properties increases the pre-computation overhead for the baseline. Note that a longer pre-computation time eventually affects the response time of a proactive auditing.

Experiments with Real Cloud Table 6.10 summarizes the obtained results. We first measure the time efficiency of LeaPS. Note that the results obtained are shorter due to the smaller size of the community cloud compared to our much larger simulated environment. Furthermore, we measure the prediction error rate (PER) of the learning tool using another dataset (for 5 days) of this cloud. For the 3.4% of PER, LeaPS affects maximum 9.62 ms additional delay in its pre-computation for the measured properties.

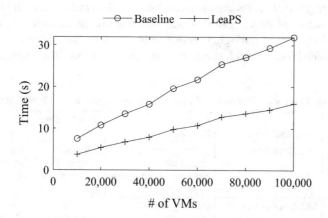

Fig. 6.25 The comparison (in seconds) of the pre-computation time between LeaPS and a baseline approach

Table 6.10 Results of experiments on real data

Properties	Learning	Pre-compute	Feedback	Verification	PER	Delay*
No bypass	7.2 s	424 ms	327 ms	5.2 ms	0.034	9.62 ms
No cross-tenant	5.97 s	419 ms	315 ms	5 ms	0.034	9.513 ms

* Note that the reported delay in LeaPS pre-computation is due to the prediction error of the learning engine

6.7   Discussion

Adapting to Other Cloud Platforms   LeaPS is designed to work with most popular cloud platforms (e.g., OpenStack [90], Amazon EC2 [3], Google GCP [40], and Microsoft Azure [71]) with a one-time effort for implementing a platform-specific interface. More specifically, LeaPS interacts with the cloud platform (e.g., while collecting logs and intercepting runtime events) through two modules: log processor and interceptor. These two modules are required to interpret implementation-specific event instances and intercept runtime events. First, to interpret platform-specific event instances as generic event types, we currently maintain a mapping of the APIs from different platforms. Table 6.11 lists some examples of such mappings. Second, the interception mechanism may require to be implemented for each cloud platform. In OpenStack, we leverage WSGI middleware to intercept and enforce the proactive auditing results so that compliance can be preserved. Through our preliminary study, we identify that almost all major platforms provide an option to intercept cloud events. In Amazon, using AWS Lambda functions, developers can write their own code to intercept and monitor events. Google GCP introduces GCP Metrics to configure charting or alerting different critical situations. Our understanding is that LeaPS can be integrated to GCP as one of the metrics similarly as the *dos_intercept_count* metric, which intends to prevent DoS attacks. The Azure Event Grid is an event managing service from Azure to monitor and control event routing, which is quite similar as our interception mechanism. Therefore, we believe

Table 6.11   API events of different cloud platforms can be mapped to LeaPS event types

LeaPS event type	OpenStack [90]	Amazon EC2-VPC [3]	Google GCP [40]	Microsoft Azure [71]
Create VM	POST/servers	aws opsworks -region create-instance	gcloud compute instances create	az vm create 1
Delete VM	DELETE/servers	aws opsworks -region delete-instance -instance-id	gcloud compute instances delete	az vm delete
Update VM	PUT/servers	aws opsworks -region update-instance -instance-id	gcloud compute instances add-tags	az vm update
Create security group	POST /v2.0/security-groups	aws ec2 create-security-group	N/A	az network nsg create
Delete security group	DELETE /v2.0/security-groups/ {security_ group_id}	aws ec2 delete-security-group -group-name {name}	N/A	az network nsg delete

Table 6.12 Existing interception supports provided by major cloud platforms

Cloud platform	Interception support
OpenStack	WSGI middleware [119]
Amazon EC2-VPC	AWS lambda function [3]
Google GCP	GCP metrics [40]
Microsoft Azure	Azure event grid [71]

that LeaPS can be an extension of the Azure Event Grid to proactively audit cloud events. Table 6.12 summarizes the interception support in these cloud platforms. The rest modules of LeaPS deal with the platform-independent data, and hence, the next steps in LeaPS are platform-agnostic.

Effects of False Positives in the Learning Technique LeaPS leverages learning techniques in a special manner so that the false positive/negative rates cannot affect the security of our system directly and rather affects the performance of our system. In LeaPS, learning parameters of Bayesian network are utilized to learn the probabilistic dependencies. Any error in learning results in a dependency model with incorrect probabilities. Later, while consulting this dependency model by the pre-computation module (as described in Sect. 6.4.2), LeaPS may choose wrong highly-likely events and perform unnecessary pre-computation for them. At the same time, LeaPS may delay the pre-computation for actual highly-likely events. The final result of such mistakes in LeaPS due to the false positives/negatives in the learning tool is the increase in the response time (as reported in Fig. 6.24).

Possibility of a DoS Attack Against LeaPS To exploit the fact that a wrong prediction may result in a delay in the LeaPS pre-computation, an attacker may conduct a DoS attack to bias the learning model step by generating fake events and hence, to exhaust LeaPS pre-computation. However, Fig. 6.24 shows that an attacker requires to inject a significantly large amount (e.g., 40% error rate) of biased event instances to cause a delay of 2 s. Moreover, biasing the model is non-trivial unless the attacker has prior knowledge of patterns of legitimate event sequences. Our future work will further investigate this possibility and its countermeasures.

Granularity of Learning The above-mentioned learning can be performed at different levels (e.g., cloud level, tenant level, and user level). The cloud level learning captures business nature only for companies using a private cloud. The tenant level learning depicts a better profile of each business or tenant. This level of learning is mainly suitable for companies strictly following process management, where users mainly follow the steps of processes. In contrast, the user level learning is suitable for smaller organizations (where no process management is followed) with fewer users (e.g., admins) who perform cloud events. Conversely, if a company follows process management, user level learning will be less useful, as different users would exhibit very similar patterns.

Dependency on Critical Event Lists The list of critical events provided by LeaPS users for each security property is very critical for LeaPS to accurately prevent any security violation. Any incompleteness in this list may result in undetected violations by LeaPS. Therefore, we provide a guideline on identifying different LeaPS inputs including lists of critical events. The steps to identify sets of event types as the inputs to the learning engine are as follows: (1) from the property definition, we identify involved cloud components; (2) we list all event types in a cloud platform involving those components; and (3) we identify the critical events (which are already provided by the tenant) from the list and further shortlist the event types based on the attack scenario. The specification of watchlist is a LeaPS input from the tenant. The specification of watchlist can be decided as follows: (1) from the property definition, the asset to keep safe is identified; (2) the objectives of a security property are to be highlighted; and (3) from the attack scenario, the parameters for the watchlist for each critical event is finalized.

Tackling Single-Step Violation The proactive auditing mechanisms fundamentally leverage the dependency in a sequence of events. In other words, proactive security auditing is mainly to detect those violations which involve multiple steps. However, there might be violations of the considered security properties with a single step. Such violations cannot be detected by the traditional steps of proactive auditing with the same response time as reported in Fig. 6.20 and may be detected by performing all steps at a single point in several seconds (e.g., around 6 s for a decent-sized cloud with 60,000 VMs as shown in Fig. 6.22), which is still faster than any other existing works (which respond in minutes). However, this response time might not be very practical. To reduce the response time or at least not to cause any significant delay, we perform a preliminary study as follows. Our initial results conducted in the testbed cloud show that OpenStack takes more than 6 s to perform almost all user requests, which implies the possibility of not resulting in any additional delay by LeaPS even for a single-step violation. Additionally, during our case studies, we observed that OpenStack performs several internal tasks to complete a user request. We may leverage this sequence of system events corresponding to a single user request to proactively perform the LeaPS steps. We elaborate those two ways of tackling single-step violations in our future work.

The Concept of Proactive Security Auditing for Clouds The concept of proactive security auditing for clouds is different than the traditional security auditing concept. Apart from ours, proactive security auditing for clouds is also proposed in [13]. Additionally, the Cloud Security Alliance (CSA) recommends continuous auditing as the highest level of auditing [19], from which our work is inspired. The current proactive and runtime auditing mechanisms are more of a combination of traditional auditing and incident management. For example, in LeaPS, we learn from incidents and intercepted events to process or detect in a similar manner as a traditional incident management system. At the same time, LeaPS verifies and enforces compliance against different security properties, which are mostly taken from different security standards, and provide detailed evidence for any violation through the LeaPS dashboard. Therefore, the concept of proactive security auditing is a combination of incident management and security auditing.

6.8   Conclusion

In this work, we proposed an automatic learning-based proactive security auditing system, *LeaPS* , which completes a mandatory pre-requisite step (e.g., log processing) and addresses the limitations of existing proactive solutions (by automating the dependency learning). To this end, we first conducted a case study on real-world cloud logs and highlighted the challenges in processing such logs. Then, we designed and implemented a log processing approach for OpenStack to feed its outputs to the learning tools to capture dependencies. Afterwards, we leveraged learning techniques (e.g., Bayesian network) to learn probabilistic dependencies for the dependency model. Finally, using such dependency models, we perform proactive security auditing. Our proposed solution is integrated to OpenStack, a popular cloud management platform. The results of our extensive experiments with both real and synthetic data show that LeaPS can be scalable for a decent-size cloud (e.g., 6 ms to audit a cloud of 100,000 VMs) and a significant improvement (e.g., about 50% faster) over existing proactive approaches. In addition, we demonstrated that other learning techniques such as sequence pattern mining algorithms can be executed on the outputs of our log processor efficiently (e.g., 18 ms to find frequencies of all possible patterns using PrefixSpan). However, this policy enforcement approach is too specific to LeaPS. In the next chapter, we generalize this method to support various runtime policy enforcement solutions.

Chapter 7
Runtime Security Policy Enforcement in Clouds

In this chapter, we explain the design and implementation of a middleware, namely *PERMON*, to apply the proactive approach to OpenStack [90], which is one of the most popular cloud platforms. The middleware is designed to intercept the attributes of user-issued requests on their path to an intended API service, and identify the requested event types based on the examined attributes. Having processed the selected parameters coupled with the identified event types, the middleware enforces the verification result by either granting or rejecting the user request. Specifically, the contributions of the following sections are:

- We present the application of our proactive verification approach to OpenStack as an efficient middleware solution, *PERMON*, for enforcing security policies at runtime in clouds.
- We detail the implementation of our approach as a Web Server Gateway Interface (WSGI) pluggable middleware for OpenStack.
- We demonstrate the usefulness of our middleware through describing a use case.
- We discuss how we address the key challenge of event type identification and the additional benefit of our solution to log processing.

7.1 Preliminaries

In this section, we provide a quick review of our proactive verification approach [66, 68] to facilitate further discussions.

Proactive Security Verification
PERMON middleware leverages our previous proactive verification approach, LeaPS [66, 68]. The main idea of LeaPS is to prepare for the verification tasks in advance, before the occurrence of actual events. Such preparation is possible due to the probable temporal relationships between user events, e.g., the occurrence of

© Springer Nature Switzerland AG 2019
S. Majumdar et al., *Cloud Security Auditing*, Advances in Information Security 76,
https://doi.org/10.1007/978-3-030-23128-6_7

creating a VM prior to attaching security groups to the same VM, which can be automatically extracted from event logs using machine learning techniques [68]. The results of such preparation (i.e., the pre-computed verification results for possible next events) are stored in a table. Therefore, the actual verification that is performed after events are received can be done by a simple lookup inside the table, which takes far less time than the actual verification does in previous approaches.

Figure 7.1 illustrates the idea behind LeaPS by comparing the way user requests are processed under a traditional intercept-and-check approach and under our proactive solution, respectively. In the upper timeline, an intercept-and-check approach intercepts and verifies the *update port* user request against the prespecified security property "no bypass" for the anti-spoofing mechanisms in the cloud, which can be violated by real-world vulnerabilities, e.g., OSSA-2015-018 [86]. A traditional intercept-and-check approach (e.g., [13]) would cause an unacceptable long delay of several minutes in medium size clouds to determine whether the request should be granted or denied.

In contrast, as depicted in the lower timeline, the proactive approach, LeaPS, works in a different way: It proactively conducts a set of pre-computations distributed among N-steps ahead of the actual occurrence of the critical operation (*update port*). The pre-computation is based on a dependency model which captures the probable temporal order between user events. This model may be manually created based on structural dependencies inherent to the cloud platform [66], e.g., a security group can only be created after the VM is created. The model can also be automatically established through applying machine learning techniques to extract frequent patterns or sequences of events from the logs, which may correspond to not only aforementioned cloud platform-specific structural dependencies but also other runtime dependencies, e.g., those due to business rules or user habits [68].

The pre-computations incrementally prepare the needed information for an efficient verification of the future critical operation. Specifically, the verification results of potential future events are pre-computed and stored in a so-called

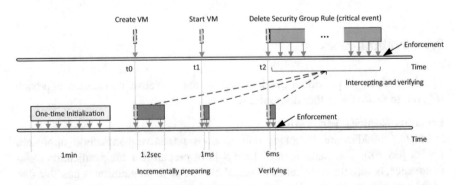

Fig. 7.1 A comparison between the delay imposed by intercept-and-check (upper) and proactive (lower) security verification approaches [66]

watchlist, and consequently the actual verification only takes negligible time (e.g., 6 ms [66]) after the actual events are received. In this chapter, we explain how our designed and implemented middleware, *PERMON*, leverages LeaPS as its proactive verification engine. We will detail the faced challenges in the design of *PERMON* to work with LeaPS, and provide solutions in later sections.

7.2 Design

In this section, we provide the high level design of *PERMON*, a proactive security policy enforcement middleware for OpenStack. The main objective of the middleware is to intercept and verify each user request against given security policies and properties such that the request is either denied or allowed to reach the intended service. We will leverage our proactive verification component LeaPS (see Sect. 7.1), although *PERMON* is designed to work with any other solution which can perform security verification. Figure 7.2 provides a high level overview about how *PERMON* works. The process consists of four main steps: Users' requests are passed through *PERMON* on their paths to the intended services *(step 1)*. *PERMON* intercepts the requests and extracts parameters which are pre-determined according to the mechanism of the deployed security solution (the proactive verification component LeaPS in our case) *(step 2)*. These parameters are verified by the security solution (LeaPS) *(step 3)*. The verification result on the legitimacy of the requested action is put into effect by *PERMON* *(step 4)*.

The proactive aspect of *PERMON* works as follows. The main objective is to pre-compute the conditions to be verified at the moment when the so-called critical events (i.e., the events that may directly cause breaches to given security properties) are received. The critical events and those events that are likely to precede the critical events have been pre-defined in the dependency model in advance. At runtime,

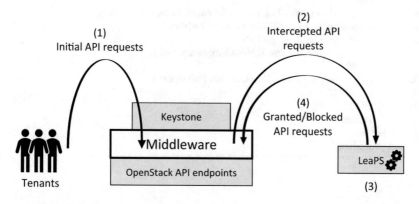

Fig. 7.2 *PERMON* overview

once *PERMON* intercepts any of those latter events, the verification module will be triggered to pre-compute the required conditions determined by the imposed security property. When the actual critical event is intercepted by *PERMON*, those pre-computed conditions are verified through simple table lookups by analyzing the intercepted attributes, and the legitimacy of the critical event is decided and enforced by *PERMON*.

The main challenges of implementing the design of *PERMON* include intercepting user requests, identifying types of events, interacting with the verification module, and enforcing the received verification results, which will be detailed in the next section.

7.3   Implementation

In this section, we detail *PERMON* implementation and its integration to Open-Stack. Figure 7.3 shows the architecture of *PERMON* with examples of inputs to various steps performed by *PERMON*. Algorithm 5 describes those steps in more details. In the following, we elaborate on the implementation of our approach employing the general mechanism of a Web Server Gateway Interface (WSGI) for implementing the middleware functionality.

Algorithm 5 *PERMON* runtime policy enforcement (*type-rules*, *field-check-rules*, *REST APIs*)

 1: **procedure** RUNTIMEPOLICYENFORCEMENT(*REST APIs*)
 2:     **for** each *call$_i$* ∈ *REST APIs* **do**
 3:         *env*[] = environFetcher(*call$_i$*)
 4:         *eventToWatch* = typeFinder(*env*[], *type-rules*)
 5:         **if** *eventToWatch* is NULL **then**
 6:             *response* = passToNextComponent(*call$_i$*)
 7:         **else**
 8:             *fields* = readFields(*env*, *field-check-rules*)
 9:             *Decision* = LeaPS(*fields*, *eventToWatch*)
10:             **if** *Decision* is Allow **then**
11:                 *response* = passToNextComponent(*call$_i$*)
12:             **else**
13:                 *response* = callBlocker("403 Forbidden")
14:             **end if**
15:         **end if**
16:         **return** *response*
17:     **end for**
18: **end procedure**

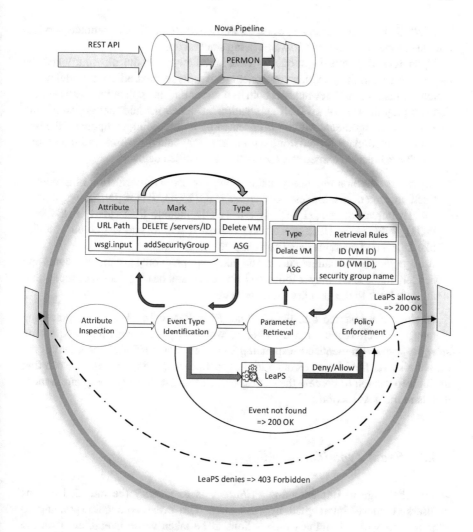

Fig. 7.3 *PERMON* architecture (URL PATH represents the conjunction of METHOD and PATH_INFO of events; the event type corresponding to adding a VM to a security group is shown as ASG)

7.3.1 *Implementing* **PERMON** *as a WSGI Middleware*

Many features of OpenStack have been implemented as pluggable middlewares based on the Web Server Gateway Interface (WSGI) standard. The main advantage to this approach is that the pluggable nature of deployed middlewares provides developers with flexibility to extend existing functionality without undertaking the cost of customizing existing codes. Following this approach, *PERMON* is designed as a WSGI middleware and is injected into Nova (OpenStack compute service) pipeline alongside other middlewares being stacked up.

Upon receiving clients' requests, each middleware in the pipeline takes the request sent from the previous component, and calls its next adjacent middleware, which means each middleware acts both as a server and an application. Any element that can play the role of a WSGI application has a call method that is passed with two positional arguments when it is executed by its preceding component; the first is a python dictionary for environment information, called *environ*, and the second is a callback function named *start_response*, as detailed below.

- *environ* is a python dictionary that contains contextual information of a request. Such information includes: REQUEST_METHOD, PATH_INFO, wsgi.input, etc. Examples of the REQUEST_METHOD include PUT, GET, POST, etc. PATH_INFO is the URL path an API request is sent to, and wsgi.input is a file-like object and contains data that is sent to the server.
- *start_response* is a callable itself, and takes two positional arguments, status and header; status is an integer followed by a string, and header is a list of tuples that are sent back to the client sending the request.

Having been inserted into the pipeline, *PERMON* is invoked by its preceding element for each request made to Nova, which is passed with the *environ* and *start_response* arguments corresponding to the request. *PERMON* here serves the role of a server for its next element in the pipeline. Inspecting *environ*, we can find useful information to process the corresponding request as we will explain in more details in the next section.

7.3.2 Event Type Identification

A key challenge in implementing *PERMON* is to identify the intended actions in users' requests intercepted by *PERMON*. The event type corresponding to such actions will determine which actions to be taken while interacting with the proactive verification module. By analyzing OpenStack API documentation, some API requests can be uniquely identified through the conjunction of their METHOD and PATH_INFO attributes. For instance, a given request can be recognized as to delete an instance if and only if its corresponding METHOD and PATH_INFO are DELETE and /servers/*server_id*, respectively.

However, not all API requests can be uniquely identified through the conjunction of their METHOD and PATH_INFO attributes. For example, as it is shown in Table 7.1, the issued requests for starting, stopping, and migrating a given server all have the same METHOD and PATH_INFO environmental variables according to API documentation. Therefore, we need another differentiating attribute to fingerprint this group of requests. By studying API documentation, we find that the METHOD field of all these requests is POST. As the data is sent in wsgi.input file-like environmental variable for requests with PUT and POST as their METHOD,

Table 7.1 Some event types can be uniquely identified using their associated API wsgi.input attributes

Event type	METHOD	PATH_INFO	wsgi.input content
Start VM	POST	/servers/*server_id*/action	"{"os-start": null}"
Stop VM	POST	/servers/*server_id*/action	"{"os-stop": null}"
Migrate VM	POST	/servers/*server_id*/action	"{"migrate": null}"

we use its content as the differentiating attribute. For example, the content of the wsgi.input object for the mentioned three requests (for starting, stopping, and migrating a given server) is {"os-start": null}, {"os-stop": null}, and {"migrate": null}, respectively (Table 7.1), which is intercepted and parsed by *PERMON* as the differentiating attribute.

By parsing the body of requests together with the combination of METHOD and PATH_INFO attributes, PERMON can map them to the type of event they are issued for. This is performed through knowing the collection of differentiating attributes of a given request and a rule mapping these attributes to a pre-defined type of event. To that end, we have embedded inside *PERMON* a set of mapping rules which associate the name of the differentiating attributes and their content with the corresponding event types. For example, the wsgi.input attribute for attaching a VM to security group SG1 is {"addSecurityGroup": {"name":"SG1"}}. To investigate whether a received request is issued to add a VM to a security group, *PERMON* examines wsgi.input attribute, and verifies its starting characters.

As mentioned earlier, we leverage our proactive verification approach, LeaPS, which triggers pre-computation at the occurrence of certain events that are determined to be more likely to be followed by a critical event. Therefore, our mapping rules for event type identification are designed to map the request attributes to those two types of events. When the former type of events are intercepted, *PERMON* will invoke LeaPS for pre-computation, and when the latter type of events are intercepted, *PERMON* will invoke LeaPS for performing the verification.

7.3.3 Interacting with LeaPS and Enforcing the Decisions

Depending on the event types, *PERMON* will invoke LeaPS either for pre-computation or for verification. During such interaction, in addition to the event types, other parameters of the corresponding event need to be extracted and passed to LeaPS in order to verify the legitimacy of the request or to pre-compute conditions. The exact parameters needed are determined based on the security properties and event types. For example, verifying requests against a security property may require the security group name and the VM ID for the event type *add security group*. Therefore, we have implemented inside *PERMON* the mappings between event types and rules to extract required parameters from the intercepted environmental variables of the corresponding request. More specifically, the extraction rules will determine which attributes to be fed into LeaPS, and the way they should be pre-processed by *PERMON*.

Finally, *PERMON* enforces the verification results of LeaPS over the legitimacy of intercepted requests. If LeaPS allows the event, *PERMON* calls the next component in the pipeline and passes to it the intercepted arguments for further processing. If LeaPS indicates denial of the request, *PERMON* blocks the request by controlling the *start_response* function as follows. One of the positional arguments, status, is fed into the callback function *start_response*. The value of this status argument of a request that has passed successfully through the previous component in the pipeline will start with a special value *2xx*. By changing this argument value to *403 Forbidden*, *PERMON* essentially blocks the request and an error message will be sent back to the user.

7.4   Use Case

In this section, we demonstrate the usefulness of *PERMON* through an example use case. We assume users' requests are made from OpenStack console. The following illustrates the functionalities of *PERMON*, which is to intercept the requests and their parameters, to interact with the verification module LeaPS, and to allow or block the corresponding requests according to the verification result.

The security property to be enforced in the use case is *no downgrade of security group for a running VM*, which prevents attackers from changing the security group of a VM, e.g., from *no_connection* to *essential* (the latter is supposedly a less restricted security group which will allow more connections to the VM, downgrading its security level). For each security group, the verification module LeaPS is initialized with security groups with higher restriction levels. The following is a sample sequence of operations illustrating how our middleware works to enforce the security property.

Stage 1 The first considered user-initiated request is creating a VM, namely test_vm.

Stage 2 User requests for attaching the VM to security group *no_connection*. Security group *no_connection* hypothetically implements the most restricted security policy.

The request passes through Nova pipeline and is sent to OpenStack networking service (Neutron) to be executed. Figure 7.4 shows the corresponding logged request made to Neutron and logged port update event for attaching security group, which means the request has reached Neutron service.

Figure 7.5 shows logged response to the underlined API request, which is associated with attaching to security group *no_connection*. The status_code 202 shows the request has been accepted and successfully processed.

```
2018-11-16 12:56:18.316 6084 INFO neutron.wsgi [req-c16f5eab-b585-4050-
9b63-ac037ea2498f f51c874fbe1d4f0c9ce46f60a1e06454
a6627ffa0c4f4a3ebaefe05c0b93f4c6 - - -] 10.0.0.101 - - [16/Nov/2018
12:56:18] "GET /v2.0/security-
groups.json?fields=id&name=no_connection&tenant_id=a6627ffa0c4f4a3ebaefe05
c0b93f4c6 HTTP/1.1" 200 282 0.082553
2018-11-16 12:56:18.394 6084 INFO symcpe_neutron.selfservice.middleware
[req-d65e89fb-a642-4633-a88a-d4b369079d56 f51c874fbe1d4f0c9ce46f60a1e06454
a6627ffa0c4f4a3ebaefe05c0b93f4c6 - - -] req_body: {"port":
{"security_groups": ["4a22ec85-a9c5-476c-a3c4-c7565bbc49d3"]}}
2018-11-16 12:56:18.820 6084 INFO neutron.wsgi [req-d65e89fb-a642-4633-
a88a-d4b369079d56 f51c874fbe1d4f0c9ce46f60a1e06454
a6627ffa0c4f4a3ebaefe05c0b93f4c6 - - -] 10.0.0.101 - - [16/Nov/2018
12:56:18] "PUT /v2.0/ports/5477c8b3-1c12-4497-a18b-3a67c6e1c618.json HTTP/
1.1" 200 1124 0.432304
```

Fig. 7.4 Logged Neutron events having triggered by receiving the request from Nova service

```
2018-11-16 12:56:18.826 5732 INFO nova.osapi_compute.wsgi.server [req-
5b1d9e5e-bf5b-40e0-a490-9554299ca1e0 f51c874fbe1d4f0c9ce46f60a1e06454
a6627ffa0c4f4a3ebaefe05c0b93f4c6 - - -] 10.0.0.101 "POST /v2.1
a6627ffa0c4f4a3ebaefe05c0b93f4c6/servers/eff8fe2d-213b-459b-8fd9-
7f5319db6457/action HTTP/1.1" status: 202 len: 273 time: 0.7806611
```

Fig. 7.5 The API response, associated with attaching test_vm to security group *no_connection*, is logged with status_code 202

```
2018-11-16 16:20:40.043 5343 INFO symcpe_nova.selfservice.middleware [req-
a0f6304c-7826-4563-93b4-41547861b479 f51c874fbe1d4f0c9ce46f60a1e06454
a6627ffa0c4f4a3ebaefe05c0b93f4c6 - - -] req_body: {"os-start": null}
```

Fig. 7.6 The intercepted body of the API request for starting a VM

Stage 3 The cloud user starts the VM. *PERMON* intercepts the request while it is passing through the pipeline and inspects its body. The request body is the content of wsgi.input of *environ* argument that is sent to the server. If *PERMON* finds the body is associated with starting a VM through verifying wsgi.input attribute of a passing request, it extracts the ID of the VM from another attribute of that request, PATH_INFO. Next, with the extracted ID, LeaPS queries the Neutron database for the security groups test_vm is attached to. According to these currently attached security groups, it populates the watchlist with the allowed security groups for the started VM.

Figure 7.6 illustrates the content of the intercepted parameter, wsgi.input. The format of PATH_INFO and the content of the request corresponding to different server actions are interpreted according to the API documentation as follows. PATH_INFO of a request for starting a VM is in the format of */servers/server_id/action*. The format provided in the API documentation is used to identify

requested actions and to extract different parameters, e.g., server_id. As it is shown
in Fig. 7.7, we keep track of the running VM with an index to the security groups it
can be legitimately attached to.

Stage 4 At this point, the attacker tries to downgrade the security group of the
running VM, test_vm, by attaching it to a less restricted security group, namely
essential (Fig. 7.8).

The content of the request is inspected by *PERMON*. Having found the body
corresponding to a security group attachment, *PERMON* extracts the security group
name and the ID of the VM, which LeaPS looks for and finds among IDs of running
VMs in its database. This means that test_vm can only be attached to more restricted
security groups than what it is currently attached to, i.e., *no_connection*. LeaPS
goes through the legitimate security groups indexed by test_vm. As security group
essential is not among them, LeaPS decides on denial of the request.

Consequently, *PERMON* refuses to pass this request to the next element in the
pipeline, and prepares its own response with status *403 Forbidden* to be sent back
to the client component (Fig. 7.5). Figure 7.9 shows the logged request body for
attaching test_vm to security group *essential*. The corresponding logged response
is shown in Fig. 7.10 with *status_code* 403, as opposed to 202 in step 2, when the

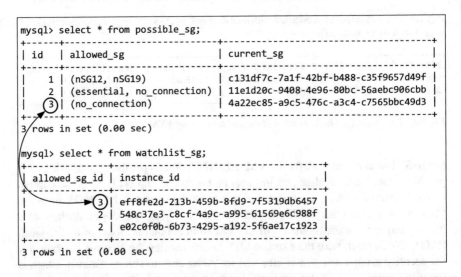

Fig. 7.7 Running VMs' IDs with an index to their allowed security groups are stored in LeaPS
database

```
adv@controller:~/OpenStackEnv$ nova add-secgroup test_vm essential
ERROR (Forbidden): Forbidden (HTTP 403) (Request-ID: req-8f4e7ebb-9057-
47cd-89e3-3b99d7c08114)
```

Fig. 7.8 Downgrading the security group of the running VM is blocked by the middleware

```
2018-11-16 22:17:20.803 19737 INFO symcpe_nova.selfservice.middleware
[req-8f4e7ebb-9057-47cd-89e3-3b99d7c08114 f51c874fbe1d4f0c9ce46f60a1e06454
a6627ffa0c4f4a3ebaefe05c0b93f4c6 - - -] req_body: {"addSecurityGroup":
{"name": "essential"}}
```

Fig. 7.9 The API request body for attaching test_vm to security group *essential* is intercepted and logged by PERMON

```
2018-11-16 22:17:20.843 19737 INFO nova.osapi_compute.wsgi.server [req-
8f4e7ebb-9057-47cd-89e3-3b99d7c08114 f51c874fbe1d4f0c9ce46f60a1e06454
a6627ffa0c4f4a3ebaefe05c0b93f4c6 - - -] 10.0.0.101 "POST /v2.1/
a6627ffa0c4f4a3ebaefe05c0b93f4c6/servers/eff8fe2d-213b-459b-8fd9-
7f5319db6457/action HTTP/1.1" status: 403 len: 226 time: 0.0461481
```

Fig. 7.10 The API response, related to attaching test_vm to security group *essential*, is logged with *status_code* 403

request was accepted and successfully processed. The common request ID between this log entry and the one shown in Fig. 7.9 shows this response corresponds to the request with the content {"addSecurityGroup": {"name":"essential"}}.

Furthermore, as opposed to step 2, no log corresponding to attaching security group *essential* can be found in logs of Neutron Service, working synchronized with Nova service, in a time frame close to the same time when the request has been made to Nova, which shows the request has been blocked by our middleware.

7.5 Discussion

Benefit to Log Analysis In addition to security policy enforcement, *PERMON* can also benefit log processing. In general, OpenStack services only log their responses to the received requests. For log analyses which aim at identifying the triggering requests, those logged responses may not provide sufficient information. Specifically, the log entries can be parsed to extract different fields among which METHOD and PATH_INFO are supposed to map to different types of requested events. However, as mentioned earlier, many events cannot be distinguished from each other. For example, all requests for invoking a server to take some specific action will have the same METHOD and PATH_INFO, which renders log interpretation infeasible. Table 7.2 shows three examples of logged responses to such requests. Except for the VM ID at the end of PATH_INFO, which is not useful for identifying the type of requested event, all three log entries have the same combination of METHOD and PATH_INTO. To this end, *PERMON* logs and examines the intercepted body of requests in order to identify the type of logged requests. This provides a feasible solution for more accurate log processing.

Table 7.2 Some logged responses to different requested event types are not distinguishable from each other

Event name	OpenStack log entry
Add security group	"POST /v2.1/b4537ffa0c4f4a3ebaefe05c0b93f4c6/servers/4c886192-43ad-4f98-90dd-34e24c84fcd0/"
Stop VM	"POST /v2.1/b4537ffa0c4f4a3ebaefe05c0b93f4c6/servers/f6128951-0c48-4a11-8b8b-5e96da77b698/"
Start VM	"POST /v2.1/b4537ffa0c4f4a3ebaefe05c0b93f4c6/servers/1223d052-bc35-485a-9237-1830bca80fd7/"

Table 7.3 The intercepted wsgi.input attributes of requests for the same event type vary depending on whether the request is issued from OpenStack command line or dashboard

	Documented mapping	wsgi.input	METHOD
Dashboard	Update VM	{"server": {"name": "VM1"}}	PUT
Command line	Add security group	{"addSecurityGroup":{"name":"SG1"}}	POST

Discrepancy Between Console and GUI Requests By studying intercepted attributes, we note that wsgi.input and METHOD of an issued request can be different depending on whether the request is made from command line or inside the OpenStack dashboard. For instance, Table 7.3 shows the content of two intercepted attributes logged by *PERMON* when a request for attaching VM1 to security group SG1 is made from command line as opposed to when it is issued from OpenStack dashboard. The logged attributes are compatible with what is indicated in the API documentation for attaching a VM to a security group in the former case; however, they are mapped to API calls for updating a VM indicated in the API documentation in the latter case. In this work, we set our matching rules according to the API documentation and we expect such discrepancy between internally generated parameters corresponding to identical requests to be addressed in future OpenStack releases.

7.6 Conclusion

This chapter presented a security policy enforcement middleware, *PERMON*, which was designed as a pluggable module in OpenStack Nova service. *PERMON* provided control over user-initiated requests according to given security policies or properties. Working along with our proactive verification module LeaPS, *PERMON* could make decisions about either allowing or denying a request in an efficient manner with only negligible delay to legitimate users. Furthermore, by inspecting the request body to identify requests that are otherwise not distinguishable, *PERMON* could bring added value to log analysis in OpenStack.

Chapter 8
Conclusion

The ever-changing and self-service nature of clouds bring the necessity to audit the cloud to ensure security compliance, which is essential for cloud providers' accountability and transparency towards their tenants. To this end, there exist three types of cloud-specific security auditing approaches: retroactive, intercept-and-check, and proactive. However, the existing works under these approaches suffer from various limitations such as failing to bridge the high-level security properties and low-level cloud configurations and logs, provide a practical response time due to the sheer scale of the cloud, or requiring manual inputs from the users which may be impractical in most cloud environments. In summary, there is a need of an automated security auditing process which can potentially overcome all these limitations.

In this book, we presented basic to advanced auditing mechanisms which overcome existing challenges/limitations and offer automation of security auditing in the cloud. To this end, we first reviewed the literature and enumerated existing challenges in automating the cloud security auditing. Then, we presented a retroactive auditing framework to verify compliance of the virtual infrastructure from the structural point of view. Afterwards, we extended that framework to audit consistent isolation between virtual networks in OpenStack-managed cloud spanning over overlay and layer 2 by considering both cloud layers' views. Furthermore, we presented a runtime security auditing framework on the cloud user level including common access control and authentication mechanisms. Moreover, we provided a proactive security auditing framework for clouds and discussed a technique to enable runtime security policy enforcement.

More specifically, in Chap. 3, we discussed an automated framework that allows auditing virtualization-related structural properties and consistency between multiple control layers. We also described the integration of our auditing system into OpenStack, one of the most used cloud infrastructure management systems.

S. Majumdar et al., *Cloud Security Auditing*, Advances in Information Security 76, https://doi.org/10.1007/978-3-030-23128-6_8

In addition, we showed the scalability and validity of our framework through our experimental results on assessing several properties related to auditing inter-layer consistency, virtual machines co-residence, and virtual resources isolation. We further extended this framework to support auditing consistent isolation between virtual networks in OpenStack-managed cloud spanning over overlay and layer 2 by considering both cloud layers' views in Chap. 4. Additionally, we integrated our auditing system into OpenStack, and presented our experimental results on assessing several properties related to virtual network isolation and consistency. Those results showed that our approach can be successfully used to detect virtual network isolation breaches for large OpenStack-based data centers in reasonable time.

To enable continuous auditing, we presented a runtime security auditing framework for the cloud with special focus on the user level including common access control and authentication mechanisms (e.g., RBAC, ABAC, and SSO) in Chap. 5. In this approach, to reduce the response time, the main idea was to perform the costly operations for only once, which is followed by significantly more efficient incremental runtime verification. We described the implementation details and evaluated the framework based on OpenStack, a widely deployed cloud management system. Furthermore, we described a learning-based proactive security auditing system in Chap. 6. To this end, we first provided our design specifics about a stand-alone log processor for clouds, which prepared raw logs for learning techniques. Consequently, we leveraged the log processor outputs to extract probabilistic dependencies from runtime events for the dependency models. Finally, through these dependency models, we proactively prepared for security critical events and prevented security violations resulting from those critical events. Finally, we discussed the potentiality to adapt the proactive middleware as a pluggable interface to OpenStack for intercepting and verifying the legitimacy of user requests at runtime while leveraging our previous work on proactive security verification to improve the efficiency in Chap. 7. We described detailed implementation of the middleware and demonstrated its usefulness through a use case.

However, the cloud security auditing approaches discussed in this book still have a few limitations, which can be addressed in future works. First, these auditing approaches are signature-based and hence cannot detect any zero-day security violations. Potentially, an anomaly based or hybrid (i.e., combining signature and anomaly based) auditing approach may improve the coverage of our solution, which we consider as a future work. Second, the efficiency of the current runtime approaches may be affected when multiple user requests appear simultaneously. A parallel or distributed approach might reduce the effect of this situation, which we consider as a potential future work. Third, the current proactive solution cannot efficiently handle single-step violations. An efficient runtime approach might help to address this concern. Fourth, our watchlist content identification process in the proactive auditing approach is currently static and manual, which may affect the accuracy of our solution. Adding a feedback loop which progressively and semi-automatically updates the content of watchlists may solve this issue. Fifth, current auditing approaches cannot prevent potential privacy concerns from the audit results. We target this problem in our future work. As a future work, we will also investigate the feasibility of applying runtime data streaming to process logs incrementally and in a more scalable manner. In addition, we intend to conduct case studies on logging of other cloud platforms to offer a platform-agnostic log processing solution, which might be useful for security solutions, which are involved with cloud logs.

References

1. A. Alimohammadifar, S. Majumdar, T. Madi, Y. Jarraya, M. Pourzandi, L. Wang, M. Debbabi, Stealthy probing-based verification (SPV): an active approach to defending software defined networks against topology poisoning attacks, in *European Symposium on Research in Computer Security* (Springer, Berlin, 2018), pp. 463–484
2. Cloud Security Alliance, The notorious nine cloud computing top threats in 2013 (2013). https://cloudsecurityalliance.org/artifacts/top-ten-big-datasecurity-and-privacy-challenges/. Accessed Jan 2019
3. Amazon, Amazon virtual private cloud. https://aws.amazon.com/vpc. Accessed 14 Feb 2018
4. Amazon Web Services, Security at scale: logging in AWS. Technical report, Amazon (2013)
5. A. Armando, R. Carbone, L. Compagna, J. Cuellar, L. Tobarra, Formal analysis of SAML 2.0 web browser single sign-on: breaking the SAML-based single sign-on for Google apps, in *Proceedings of the 6th ACM Workshop on Formal Methods in Security Engineering* (2008)
6. BayesFusion, GeNIe and SMILE. https://www.bayesfusion.com. Accessed 14 Feb 2018
7. M. Bellare, B. Yee, Forward integrity for secure audit logs. Technical report, Citeseer (1997)
8. M. Ben-Ari, *Mathematical Logic for Computer Science* (Springer, London, 2012)
9. N. Bjørner, K. Jayaraman, Checking cloud contracts in Microsoft Azure, in *Distributed Computing and Internet Technology* (Springer, Berlin, 2015)
10. S. Bleikertz, Automated security analysis of infrastructure clouds. Master's Thesis, Technical University of Denmark and Norwegian University of Science and Technology (2010)
11. S. Bleikertz, T. Groß, M. Schunter, K. Eriksson, Automated information flow analysis of virtualized infrastructures, in *European Symposium on Research in Computer Security (ESORICS)* (Springer, Berlin, 2011), pp. 392–415
12. S. Bleikertz, C. Vogel, T. Groß, Cloud Radar: near real-time detection of security failures in dynamic virtualized infrastructures, in *Proceedings of the 30th Annual Computer Security Applications Conference (ACSAC)* (ACM, New York, 2014), pp. 26–35
13. S. Bleikertz, C. Vogel, T. Groß, S. Mödersheim, Proactive security analysis of changes in virtualized infrastructures, in *Proceedings of the 31st Annual Computer Security Applications Conference (ACSAC)* (ACM, New York, 2015), pp. 51–60
14. S. Butt, H.A. Lagar-Cavilla, A. Srivastava, V. Ganapathy, Self-service cloud computing, in *Proceedings of the 2012 ACM Conference on Computer and Communications Security* (ACM, New York, 2012), pp. 253–264
15. Cloud Security Alliance, Security guidance for critical areas of focus in cloud computing v 3.0 (2011). https://cloudsecurityalliance.org/research/guidance/. Accessed Sept 2014

© Springer Nature Switzerland AG 2019

S. Majumdar et al., *Cloud Security Auditing*, Advances in Information Security 76,
https://doi.org/10.1007/978-3-030-23128-6

16. Cloud Security Alliance, Top ten big data security and privacy challenges (2012). https://cloudsecurityalliance.org/artifacts/top-ten-big-datasecurity-and-privacy-challenges/. Accessed Jan 2019
17. Cloud Security Alliance, Cloud control matrix CCM v3.0.1 (2014). https://cloudsecurityalliance.org/research/ccm/. Accessed 14 Feb 2018
18. Cloud Security Alliance, Cloud computing top threats in 2016 (2016)
19. Cloud Security Alliance, CSA STAR program and open certification framework in 2016 and beyond (2016). https://downloads.cloudsecurityalliance.org/star/csa-star-program-cert-prep.pdf. Accessed 14 Feb 2018
20. Crandall et al. Virtual networking management white paper. Technical report, DMTF (2012). DMTF Draft White Paper
21. EU Project, Certification infrastructure for multi-layer cloud services project (cumulus) (2012). http://www.cumulus-project.eu. Accessed Jan 2019
22. Data Center Knowledge, Survey: one-third of cloud users' clouds are private, heavily OpenStack (2015). http://www.datacenterknowledge.com/archives/2015/01/30/survey-half-of-private-clouds-are-openstack-clouds. 14 Feb 2018
23. V. Del Piccolo, A. Amamou, K. Haddadou, G. Pujolle, A survey of network isolation solutions for multi-tenant data centers. IEEE Commun. Surv. Tutorials **18**(4), 2787–2821 (2016)
24. A.P. Dempster, N.M. Laird, D.B. Rubin, Maximum likelihood from incomplete data via the em algorithm. J. R Stat. Soc. Ser. B **39**, 1–38 (1977)
25. M. Dhawan, R. Poddar, K. Mahajan, V. Mann, Sphinx: detecting security attacks in software-defined networks, in *NDSS Symposium*, San Diego, California. Internet Society (2015)
26. Distributed Management Task Force, Inc. Cloud auditing data federation (2016). https://www.dmtf.org/standards/cadf
27. F.H.-U. Doelitzscher, Security audit compliance for cloud computing. Ph.D. Thesis, Plymouth University (2014)
28. F. Doelitzscher, C. Fischer, D. Moskal, C. Reich, M. Knahl, N. Clarke, Validating cloud infrastructure changes by cloud audits, in *Eighth World Congress on Services (SERVICES)* (IEEE, Piscataway, 2012), pp. 377–384
29. E. Dolzhenko, J. Ligatti, S. Reddy, Modeling runtime enforcement with mandatory results automata. Int. J. Inf. Sec. **14**(1), 47–60 (2015)
30. Elasticsearch, Logstash. https://www.elastic.co/products/logstash. Accessed 14 Feb 2018
31. ENISA, European union agency for network and information security (2016). https://www.enisa.europa.eu
32. H.P. Enterprise, Hpe helion eucalyptus (2017). http://www8.hp.com/us/en/cloud/helion-eucalyptus.html
33. D.F. Ferraiolo, R. Sandhu, S. Gavrila, D.R. Kuhn, R. Chandramouli, Proposed NIST standard for role-based access control. ACM Trans. Inf. Syst. Secur. **4**(3), 224–274 (2001)
34. S.N. Foley, U. Neville, A firewall algebra for OpenStack, in *Conference on Communications and Network Security (CNS)* (IEEE, Piscataway, 2015), pp. 541–549
35. P. Fournier-Viger, SPMF, an open-source data mining library. http://www.philippe-fournier-viger.com/spmf/index.php. Accessed 14 Feb 2018
36. P. Fournier-Viger, C.-W. Wu, V.S. Tseng, Mining maximal sequential patterns without candidate maintenance, in *International Conference on Advanced Data Mining and Applications* (Springer, Berlin, 2013), pp. 169–180
37. getcloudify.org, OpenStack in numbers - the real stats (2014). http://getcloudify.org
38. N. Ghosh, D. Chatterjee, S.K. Ghosh, S.K. Das, Securing loosely-coupled collaboration in cloud environment through dynamic detection and removal of access conflicts. IEEE Trans. Cloud Comput. **4**, 1 (2014)
39. A. Gomariz, M. Campos, R. Marin, B. Goethals, Clasp: an efficient algorithm for mining frequent closed sequences, in *Pacific-Asia Conference on Knowledge Discovery and Data Mining* (Springer, Berlin, 2013), pp. 50–61
40. Google, Google cloud platform. https://cloud.google.com. Accessed 14 Feb 2018

41. A. Gouglidis, I. Mavridis, domRBAC: an access control model for modern collaborative systems. Comput. Secur. **31**, 540–556 (2012)
42. A. Gouglidis, I. Mavridis, V.C. Hu, Security policy verification for multi-domains in cloud systems. Int. J. Inf. Sec. **13**(2), 97–111 (2014)
43. T. Groß, Security analysis of the SAML single sign-on browser/artifact profile, in *Proceedings of 19th Annual Computer Security Applications Conference (ACSAC)* (2003)
44. S. Gutz, A. Story, C. Schlesinger, N. Foster, Splendid isolation: a slice abstraction for software-defined networks, in *Proceedings of the First Workshop on Hot Topics in Software Defined Networks*, HotSDN '12 (ACM, New York, 2012), pp. 79–84
45. D. Heckerman, A tutorial on learning with bayesian networks, in *Learning in Graphical Models* (Springer, Berlin, 1998), pp. 301–354
46. S. Hong, L. Xu, H. Wang, G. Gu, Poisoning network visibility in software-defined networks: new attacks and countermeasures, in *Proceedings of 2015 Annual Network and Distributed System Security Symposium (NDSS'15)* (2015)
47. V.C. Hu, D. Ferraiolo, R. Kuhn, A. Schnitzer, K. Sandlin, R. Miller, K. Scarfone, Guide to attribute based access control (ABAC) definition and considerations. NIST SP, 800 (2014)
48. IBM, Safeguarding the cloud with IBM security solutions. Technical Report, IBM Corporation (2013)
49. IBM Corporation, IBM point of view: security and cloud computing (2009)
50. Z. Ismail, C. Kiennert, J. Leneutre, L. Chen, Auditing a cloud provider's compliance with data backup requirements: a game theoretical analysis. IEEE Trans. Inf. Forensics Secur. **11**(8), 1685–1699 (2016)
51. ISO Std IEC, ISO 27002: 2005. Information technology-security techniques- code of practice for information security management. ISO (2005)
52. ISO Std IEC, ISO 27017. Information technology- security techniques- code of practice for information security controls based on ISO/IEC 27002 for cloud services (DRAFT) (2012). http://www.iso27001security.com/html/27017.html. Accessed 14 Feb 2018
53. W.A. Jansen, Inheritance properties of role hierarchies, in *21st National Information Systems Security Conference* (NISSC) (1998)
54. X. Jin, Attribute based access control model. https://blueprints.launchpad.net/keystone/%2Bspec/attribute-based-access-control
55. X. Jin, Attribute based access control and implementation in infrastructure as a service cloud. Ph.D. Thesis, The University of Texas at San Antonio (2014)
56. H. Kai, H. Chuanhe, W. Jinhai, Z. Hao, C. Xi, L. Yilong, Z. Lianzhen, W. Bin, An efficient public batch auditing protocol for data security in multi-cloud storage, in *8th ChinaGrid Annual Conference (ChinaGrid)* (IEEE, Piscataway, 2013), pp. 51–56
57. S.L. Lauritzen, The EM algorithm for graphical association models with missing data. Comput. Stat. Data Anal. **19**(2), 191–201 (1995)
58. M. Li, W. Zang, K. Bai, M. Yu, P. Liu, Mycloud: supporting user-configured privacy protection in cloud computing, in *Proceedings of the 29th Annual Computer Security Applications Conference (ACSAC)* (ACM, New York, 2013), pp. 59–68
59. J. Ligatti, L. Bauer, D. Walker, Run-time enforcement of nonsafety policies.ACM Trans. Inf. Syst. Secur. **12**(3), 19 (2009)
60. J. Ligatti, S. Reddy, A theory of runtime enforcement, with results, in *European Symposium on Research in Computer Security (ESORICS)* (Springer, Berlin, 2010), pp. 87–100
61. Z. Lu, Z. Wen, Z. Tang, R. Li, Resolution for conflicts of inter-operation in multi-domain environment. Wuhan Univ. J. Nat. Sci. **12**(5), 955–960 (2007)
62. Y. Luo, W. Luo, T. Puyang, Q. Shen, A. Ruan, Z. Wu, OpenStack security modules: a least-invasive access control framework for the cloud, in *IEEE 9th International Conference on Cloud Computing (CLOUD)* (2016)
63. T. Madi, S. Majumdar, Y. Wang, Y. Jarraya, M. Pourzandi, L. Wang, Auditing security compliance of the virtualized infrastructure in the cloud: application to OpenStack, in *Proceedings of the Sixth ACM Conference on Data and Application Security and Privacy (CODASPY)* (ACM, New York, 2016), pp. 195–206

64. T. Madi, Y. Jarraya, A. Alimohammadifar, S. Majumdar, Y. Wang, M. Pourzandi, L. Wang, M. Debbabi, ISOTOP: auditing virtual networks isolation across cloud layers in OpenStack. ACM Trans. Privacy Secur. **22**, 1 (2018)

65. S. Majumdar, T. Madi, Y. Wang, Y. Jarraya, M. Pourzandi, L. Wang, M. Debbabi, Security compliance auditing of identity and access management in the cloud: application to OpenStack, in *7th International Conference on Cloud Computing Technology and Science (CloudCom)* (IEEE, Piscataway, 2015), pp. 58–65

66. S. Majumdar, Y. Jarraya, T. Madi, A. Alimohammadifar, M. Pourzandi, L. Wang, M. Debbabi, Proactive verification of security compliance for clouds through pre-computation: application to OpenStack, in *European Symposium on Research in Computer Security (ESORICS)* (Springer, Berlin, 2016), pp. 47–66

67. S. Majumdar, Y. Jarraya, M. Oqaily, A. Alimohammadifar, M. Pourzandi, L. Wang, M. Debbabi, Leaps: learning-based proactive security auditing for clouds, in *European Symposium on Research in Computer Security (ESORICS)* (Springer, Berlin, 2017), pp. 265–285

68. S. Majumdar, Y. Jarraya, M. Oqaily, A. Alimohammadifar, M. Pourzandi, L. Wang, M. Debbabi, Leaps: learning-based proactive security auditing for clouds, in ed. by S.N. Foley, D. Gollmann, E. Snekkenes. *Computer Security – ESORICS 2017* (Springer, Cham, 2017), pp. 265–285

69. S. Majumdar, T. Madi, Y. Wang, Y. Jarraya, M. Pourzandi, L. Wang, M. Debbabi, User-level runtime security auditing for the cloud. IEEE Trans. Inf. Forensics Secur. **13**(5), 1185–1199 (2018)

70. R. Martins, V. Manquinho, I. Lynce, An overview of parallel sat solving. Constraints **17**(3), 304–347 (2012)

71. Microsoft, Microsoft Azure virtual network. https://azure.microsoft.com. Accessed 14 Feb 2018

72. Midokura, Run midonet at scale (2017). http://www.midokura.com/midonet/

73. H. Moraes, M.A.M. Vieira, I Cunha, D. Guedes, Efficient virtual network isolation in multi-tenant data centers on commodity ethernet switches, in *2016 IFIP Networking Conference (IFIP Networking) and Workshops*, Vienna (IEEE, Piscataway, 2016), pp. 100–108

74. K. Murphy, A brief introduction to graphical models and Bayesian networks (1998). https://www.cs.ubc.ca/~murphyk/Bayes/bayes_tutorial.pdf. Accessed Jan 2019

75. S. Narain, Network configuration management via model finding, in *Proceedings of the 19th Conference on Large Installation System Administration Conference (LISA)* (2005), p. 15

76. The European Network and Information Security Agency, Cloud computing benefits, risks and recommendations for information security (2012). https://resilience.enisa.europa.eu/cloud-security-andresilience/publications/cloud-computing-benefits-risks-and-recommendations-for-informationsecurity. Accessed Jan 2019

77. NIST, SP 800-53. Recommended security controls for federal information systems (2003)

78. OASIS, Security assertion markup language (SAML) (2016). http://www.oasis-open.org/committees/security

79. H.-K. Oh, S.-H. Jin, The security limitations of SSO in OpenID, in *10th International Conference on Advanced Communication Technology* (2008)

80. ONF, Openflow switch specification (2013). http://www.gesetze-im-internet.de/englisch_bdsg

81. Open Data Center Alliance, Open data center alliance usage: cloud based identity governance and auditing rev. 1.0. Technical Report, Open Data Center Alliance (2012)

82. Opendaylight, The OpenDaylight platform (2015). https://www.opendaylight.org/

83. OpenID Foundation, OpenID: the internet identity layer (2016). http://openid.net

84. OpenStack, Ossa-2014-008: routers can be cross plugged by other tenants (2014). https://security.openstack.org/ossa/OSSA-2014-008.html. Accessed 14 Feb 2018

85. OpenStack, Ossa-2014-008: routers can be cross plugged by other tenants (2014). https://security.openstack.org/ossa/OSSA-2014-008.html

86. OpenStack, Neutron firewall rules bypass through port update (2015). https://security.openstack.org/ossa/OSSA-2015-018.html

87. OpenStack, Nova network configuration allows guest VMS to connect to host services (2015). https://wiki.openstack.org/wiki/OSSN/OSSN-0018
88. OpenStack, Nova network security group changes are not applied to running instances (2015). https://security.openstack.org/ossa/OSSA-2015-021.html. Accessed 14 Feb 2018
89. OpenStack, OpenStack congress (2015). https://wiki.openstack.org/wiki/Congress. Accessed 14 Feb 2018
90. OpenStack, OpenStack open source cloud computing software (2015). http://www.openstack.org. Accessed 14 Feb 2018
91. OpenStack, OpenStack audit middleware (2016). http://docs.openstack.org/developer/keystonemiddleware/audit.html. Accessed 14 Feb 2018
92. OpenStack, OpenStack user survey (2016). https://www.openstack.org/assets/survey/October2016SurveyReport.pdf. Accessed 14 Feb 2018
93. J. Pearl, *Causality: Models, Reasoning and Inference* (Cambridge University Press, 2000)
94. J. Pei, J. Han, B. Mortazavi-Asl, J. Wang, H. Pinto, Q. Chen, U. Dayal, M.-C. Hsu, Mining sequential patterns by pattern-growth: the prefixspan approach. IEEE Trans. Knowl. Data Eng. **16**(11), 1424–1440 (2004)
95. D. Perez-Botero, J. Szefer, R.B. Lee, Characterizing hypervisor vulnerabilities in cloud computing servers, in *Proceedings of the 2013 International Workshop on Security in Cloud Computing*, Cloud Computing '13 (ACM, New York, 2013), pp. 3–10
96. D. Petcu, C. Craciun, Towards a security SLA-based cloud monitoring service, in *Proceedings of the 4th International Conference on Cloud Computing and Services Science (CLOSER)* (2014), pp. 598–603
97. B. Pfaff, J. Pettit, T. Koponen, K. Amidon, M. Casado, S. Shenker, Extending networking into the virtualization layer, in *HotNets*, YorkCity (ACM, New York, 2009), pp. 598–603
98. P. Pritzker, P.D. Gallagher, NIST cloud computing standards roadmap. Technical Report, NIST, Gaithersburg (2013). NIST Special Publication 500-291
99. N. Pustchi, R. Sandhu, MT-ABAC: a multi-tenant attribute-based access control model with tenant trust, in *Network and System Security (NSS)*(2015)
100. K. Ren, C. Wang, Q. Wang, Security challenges for the public cloud. IEEE Internet Comput. **16**(1), 69–73 (2012)
101. T. Ristenpart, E. Tromer, H. Shacham, S. Savage, Hey, you, get off of my cloud: exploring information leakage in third-party compute clouds, in *Proceedings of the 16th ACM Conference on Computer and Communications Security*, CCS '09 (ACM, New York, 2009), pp. 199–212
102. R. Sandhu, The authorization leap from rights to attributes: maturation or chaos?, in *Proceedings of the 17th ACM symposium on Access Control Models and Technologies* (2012)
103. F.B. Schneider, Enforceable security policies. Trans. Inf. Syst. Secur. **3**(1), 30–50 (2000)
104. C.S. Sean Convery, Hacking layer 2: Fun with ethernet switches (2002). BlackHat Briefings
105. R. Skowyra, L. Xu, G. Gu, T. Hobson, V. Dedhia, J. Landry, H. Okhravi, Effective topology tampering attacks and defenses in software-defined networks, in *Proceedings of the 48th Annual IEEE/IFIP International Conference on Dependable Systems and Networks (DSN'18)* (2018)
106. M. Solanas, J. Hernandez-Castro, D. Dutta, Detecting fraudulent activity in a cloud using privacy-friendly data aggregates. Technical Report, arXiv preprint (2014)
107. A. Tabiban, S. Majumdar, L. Wang, M. Debbabi, Permon: an openstack middleware for runtime security policy enforcement in clouds, in *Proceedings of the 4th IEEE Workshop on Security and Privacy in the Cloud (SPC 2018)* (2018)
108. N. Tamura, M. Banbara, Sugar: a CSP to SAT translator based on order encoding, in *Proceedings of the Second International CSP Solver Competition* (2008), pp. 65–69
109. N. Tamura, M. Banbara, Syntax of Sugar CSP description. http://bach.istc.kobe-u.ac.jp/sugar (2010)
110. B. Tang, R. Sandhu, Extending openstack access control with domain trust, in *Network and System Security* (Springer, Berlin, 2014), pp. 54–69

111. Y. Diogenes, Cloud services foundation reference architecture-reference model (2013). http://blogs.technet.com/b/cloudsolutions/archive/2013/08/15/cloudservices-foundation-reference-architecture-reference-model.aspx. Accessed Jan 2019
112. K.W. Ullah, A.S. Ahmed, J. Ylitalo, Towards building an automated security compliance tool for the cloud, in *12th International Conference on Trust, Security and Privacy in Computing and Communications (TrustCom)* (IEEE, Piscataway, 2013), pp. 1587–1593
113. VMware, VMware vCloud director. https://www.vmware.com. Accessed 14 Feb (2018)
114. O. vSwitch, Open vswitch (2016). http://openvswitch.org/
115. R. Wang, S. Chen, X. Wang, Signing me onto your accounts through Facebook and Google: a traffic-guided security study of commercially deployed single-sign-on web services. In *Proceedings of the IEEE Symposium on Security and Privacy (IEEE S&P)* (2012)
116. C. Wang, S.S. Chow, Q. Wang, K. Ren, W. Lou, Privacy-preserving public auditing for secure cloud storage. IEEE Trans. Comput. **62**(2), 362–375 (2013)
117. Y. Wang, T. Madi, S. Majumdar, Y. Jarraya, M. Pourzandi, L. Wang, M. Debbabi, Tenant-guard: scalable runtime verification of cloud-wide vm-level network isolation, in *Proceedings of 2017 Annual Network and Distributed System Security Symposium (NDSS'17)* (2017)
118. Y. Wang, Q. Wu, B. Qin, W. Shi, R.H. Deng, J. Hu, Identity-based data outsourcing with comprehensive auditing in clouds. IEEE Trans. Inf. Forensics Secur. **12**(4), 940–952 (2017)
119. WSGI, Middleware and libraries for WSGI (2016). http://wsgi.readthedocs.io/en/latest/libraries.html. Accessed 15 Feb 2018
120. H. Zeng, S. Zhang, F. Ye, V. Jeyakumar, M. Ju, J. Liu, N. McKeown, A. Vahdat, Libra: Divide and conquer to verify forwarding tables in huge networks, in *Proceedings of the 11th USENIX Symposium on Networked Systems Design and Implementation (NSDI)*, vol. 14 (2014), pp. 87–99
121. S. Zhang, S. Malik, Sat based verification of network data planes, in ed. by D. Van Hung, M. Ogawa. *Automated Technology for Verification and Analysis*. Lecture Notes in Computer Science, vol. 8172 (Springer, Cham, 2013), pp. 496–505
122. Y. Zhang, A. Juels, A. Oprea, M.K. Reiter, Homealone: co-residency detection in the cloud via side-channel analysis, in *2011 IEEE Symposium on Security and Privacy* (IEEE, Piscataway, 2011), pp. 313–328
123. X. Zhu, S. Song, J. Wang, S.Y. Philip, J. Sun, Matching heterogeneous events with patterns, in *30th International Conference on Data Engineering (ICDE)* (IEEE, Piscataway, 2014), pp. 376–387

Printed in the United States
By Bookmasters